AF182585

CREATING THE FUTURE WITH ALL FINANCE AND FINANCIAL CONGLOMERATES

CREATING THE FUTURE WITH ALL FINANCE AND FINANCIAL CONGLOMERATES

by

LUTGART VAN DEN BERGHE
The Vlerick School of Management,
University of Gent

and

KURT VERWEIRE
Erasmus University Rotterdam

SPRINGER-SCIENCE+BUSINESS MEDIA, B.V.

A C.I.P. Catalogue record for this book is available from the Library of Congress.

ISBN 978-1-4419-5043-7 ISBN 978-1-4757-4881-9 (eBook)
DOI 10.1007/978-1-4757-4881-9

Printed on acid-free paper

All Rights Reserved
© 1998 Springer Science+Business Media Dordrecht
Originally published by Kluwer Academic Publishers in 1998
No part of the material protected by this copyright notice may be reproduced or utilized in any form or by any means, electronic or mechanical, including photocopying, recording or by any information storage and retrieval system, without written permission from the copyright owner.

CONTENT

INTRODUCTION

The aim of this book is to give an in-depth view of the diversification strategies of financial institutions, financial conglomerates, the growing interface between banks, insurance companies and investment companies, as well as of the development of combined products and *all finance* or service bundles.

Since many disciplines are involved in this analysis and theoretical as well as practical relevance is our touchstone, the different parts of this book can be read as separate modules. It is only by combining all of these modules that a complete picture can be drawn.

Part A of section I is mainly looking at the theoretical fundaments of this diversification trend and this from different angles.

First a brief distinction is made between bancassurance, assurfinance, all finance and financial conglomerates A first glance is cast at the evolution from co-operative interfaces towards competitive interfaces between banks and insurance companies. Furthermore the different methods of entry into diversified territory are detailed.

Chapter 2 analyses the trend from bancassurance and assurfinance towards all finance in more detail and tries to develop a new theoretical framework for understanding the differences within this diversification trend.

In chapter 3 a thorough analysis of the ratio behind these diversification strategies is developed and the general literature on diversification and synergies is compared with the specific concepts and experiences in the financial sector. The apparent contradiction between de-merger and refocusing on the general business level with the growing diversification of banks and insurance companies is de-mystified.

The Financial Conglomerates Control Board (F.C.C.B.) is developed as a construct to categorise the many formulas in order to establish and structure a financial conglomerate. Moreover this instrument can be used to evaluate the degree of integration (being one of the main drivers for success) as well as to identify the possible risks involved.

Part B of section I tests the economic value of all finance and financial conglomerates. With the help of the F.C.C.B. in chapter 5 a comparatative qualitative analysis is made of a number of different diversification strategies.

Chapter 6 tries to quantify the overall effect of diversification strategies and tests for the relevant determinants of the performance of diversified groups versus specialised banks and insurance companies.

Part C of section I details the modern strategic recipes and investigates in chapter 7 their relevance for the financial sector. Chapter 8 then looks at some case studies that apply the strategic concepts of stretch and leverage from different perspectives, either at the structural level or at the level of the definition of the core business and the appropriate product development.

Section II focuses on the management of the risks involved in financial conglomerates and details some relevant regulatory issues and solutions. Chapter 9 makes the inventory of the main risks involved and looks at the opinion of the European Commission and other relevant supervisory authorities. Also the peculiar policy of the U.S. in relation to diversifying financial firms is briefly analysed.Chapter 10 gives an in-depth evaluation of the proposed directive on insurance groups which is an essential step towards a regulation on financial conglomerates.

Section III gives a synthesis of the major findings.

Ghent, March 1998

SECTION I

THE ACADEMIC SEARCH FOR AN UNDERSTANDING OF ALL FINANCE AND FINANCIAL CONGLOMERATES

A STRATEGIC AND ECONOMIC ANALYSIS WITHIN THE FINANCIAL SERVICES INDUSTRY

PART A

The theoretical foundations

1

SETTING THE SCENERY

WHAT'S IN A NAME?

Distinguishing between bancassurance, assurfinance, all finance and financial conglomerates

The financial services industry has witnessed considerable changes since the beginning of the eighties. One of the most remarkable trends has been the breaking down of the barriers which existed between the different parts of the financial sector. Now most banks offer a range of insurance products beside their traditional financial services portfolio. On the other hand, quite a number of insurance companies have moved in the direction towards offering financial services. Sometimes this phenomenon has been labelled *bancassurance*, others have called it *all finance* and those suppliers are often referred to as *financial conglomerates*. Sectoral specialists are certainly familiar with most of the terms mentioned. Nevertheless, it is clear that these terms may mean completely different things for different countries, companies or individuals.

They all have one thing in common: they relate to the diversification strategy of banks, insurance companies and other financial institutions. All these diversification strategies aim at cross-selling each other's products. The more sophisticated ones involve a further integration of products, marketing strategies and even production and management processes. It is therefore not only necessary to analyse the different types of financial conglomerates (and structures) but also to look more carefully at the different strategies behind the diversification in the banking and insurance sector.

First we want to differentiate the terms *bancassurance* and *assurfinance* from the concept of *all finance*, to be sure that everyone gives the same content and meaning to the business practices we will be studying further on.

Bancassurance designates the strategy of a bank to cross-sell insurance products through its own distribution channels, usually branches. Of course there are different ways to realise this strategy: the bank can produce the product itself and become the risk bearer through its own insurance company; she can realise her diversification strategy with the help of a partner in a (creative) joint venture or limit herself to a pure agent or brokerage role (respectively with one insurance partner or as an independent intermediary).

Assurfinance is the opposite diversification strategy whereby an insurance company and/or an insurance intermediary cross-sells financial products.

In both cases, a more supply-driven or institutional approach is being used, applying marketing techniques like cross-selling and multi-channel distribution.

All finance can be seen as a more product and market-oriented approach, whereby products produced in different *factories* are unbundled and rebundled to tailor them to the needs of specific client segments in order to offer them an integrated personalised solution.

In order to develop such strategies, large financial groups consisting of both banking and insurance subsidiaries, originated. These financial groups are mostly referred to as *financial conglomerates.*

The concept of financial conglomerates redefined

In a workshop, organised by the Centre of Competence on Financial Conglomerates in Rotterdam (September, 1994), we have discussed the definition issue and we came to the following conclusion (Van den Berghe, 1995):

- it relates to a group of enterprises, and;
- this group is formed by different types of financial institutions. The difference between the firms involved, is often related to the difference in their supervisory rules, which in itself must be the translation of more fundamental differences between the respective core businesses. Although the term *conglomerate* initially reflects a rather heterogeneous set of firms, the fact that we talk about financial institutions already indicates that there exists some complementarity between the respective core businesses.

Although this definition may seem quite straightforward, sometimes difficulties may arise in applying this definition. These difficulties trace back to difficulties with the distinct elements of the definition. Different criteria may be used to determine whether a company belongs to a group. In chapter 10 we will analyse the question whether a legal entity, called *group,* can be the reference base to define a financial conglomerate from a regulatory perspective. For the strategic analysis of the diversified financial firm another approach is necessary.

This confirms the research results of our previous publication where we concluded that:

> "definitions often differ according to the type of the user(s) or the use made of it. For instance, the viewpoint of a shareholder might be significantly different from that of a creditor or a supervisor" (Van den Berghe, 1995: 2).

For the purpose of strategic and economic analysis it is therefore necessary to make the distinction between *financial conglomerates sensu stricto* and *financial conglomerates sensu lato*. In the former case the members of the group are linked through qualifying holdings and have a common strategy and mutual responsibilities. A qualifying holding indicates that one company has the possibility to influence the management of another company. Financial conglomerates, in the broadest sense, are not linked through qualifying holdings but they have marketing joint ventures with common trade names or distribution agreements so that the entrepreneurial risks are combined. One of the conditions necessary is the lasting character of the co-operation and the interactions between the business activities involved.

HOW IT ALL STARTED

Co-operative interfaces between banks and insurance companies

The formation of financial conglomerates does, however, not coincide completely with the growing market interface between insurance companies, banks and investment firms. Belgium for instance has been a country of financial holdings which did not operate in the market as financial conglomerates for quite a long time. On the other hand, even before such large financial groups were formed, there were important operational relations between different players in the financial market.

According to a study of Maycock (1986) the financial conglomerate movement mainly started with banks who elaborated separate subsidiaries for a wide range of financial services such as leasing, consumer finance, mortgage, financial engineering, unit trust management, insurance broking and consultancy services.

The diversification of the banking sector was driven by different forces. One of the most significant forces has been the desintermediation in commercial banking. The traditional function of a credit institution is to act as an intermediary between economic agents with financial surpluses and agents with financial shortages. During the former decade, some structural trends have evolved, both on the assets and the liabilities side of the bank balance sheet.

On the assets side, credit institutions are confronted with the replacement of straight bank loans by market-determined sources of financing. For example large corporations with a good rating prefer the issuance of bonds and other securities to get the necessary capital. On the liabilities side, there is a substantial outflow of deposits to a wide range of new financial products offered by companies of different sectors.

In addition, banks faced new capital adequacy rules which put further pressure on the banks' profitability. Examples of significant moves in this respect are the increase of off-balance sheet transactions (e.g. securitisation) and increased fee-based business (e.g. financial engineering, investment advice and insurance selling).

Therefore, banks have looked for new sources of profit, preferably without increasing their necessary solvency level. All fee income was therefore a welcome source of diversification. By acting as agents (or brokers) for insurance companies, banks could increase their fee income and enlarge their service assortment with some *high touch* products. To a certain extent, also insurance companies benefited from this development: they had an additional distribution channel at their disposal. This agency role banks play for insurance companies has a long history. Dickinson and Dinenis (1993) give several reasons why this has happened:

> "It arises in part because of the complementary nature of banking and insurance products, and in part because both banks and insurance companies can reduce the search costs for their existing customers wishing to extend the range of financial products they purchase."

It is remarkable that at that time, insurance companies did not show the same intensity to diversify into non-core business activities.

Competitive interfaces between banks and insurance companies

The situation changed when banks decided to set up their own insurance companies: from this moment on banks were considered as competitors for insurance companies.

Lafferty Business Research (1991) sums up the factors that have prompted the banks to do so, and which created an entirely new environment for the financial services sector:

- a dramatic alternation in demographic patterns;
- a shift in the emphasis of consumer demands;
- corresponding changing savings trends.

In the 1980s the population growth rates have significantly slowed down whereas the average lifetime has increased significantly. The combination of these two trends leads to a *greying* population. This demographic pattern has serious consequences on the overall economic environment, and especially on the provision of social security. The increasing proportion of the elderly relative to the working population is placing a strain on public pension systems. These systems generally work through a redistribution scheme and as a result, the financing base comes under pressure (a decreasing number of contributors) while at the same time the financing needs increase drastically (a growing number of beneficiaries who live longer and have earned higher wages). This tendency is putting a severe strain on governments trying to maintain state pension levels.

For the higher-income groups the public provisions are generally not enough to ensure the same standard of living. As a consequence, people try to set up their own additional pension scheme. Given the financial pressure on the state pension systems and the drive towards decreasing the government deficits the level of public pension could well decrease in the (near) future. Consequently, also the lower income levels increasingly feel the need for extra private pension savings.

Thanks to the growing prosperity and the expectations of a higher standard of living, the budgetary margin for pension savings increased, creating a growing emphasis on long-term, high-yield financial savings (Lafferty Business Research, 1991). This partly explains the success of alternative investment vehicles such as unit-linked and universal life products.

Life insurance in general has been a booming industry since the eighties. Of course the retirement issue we just mentioned, has certainly had a positive impact on the growth of the life insurance sector, but other factors have been at work as well. The level of inflation is another determining factor for the economic potential of life insurance. Due to the discipline imposed by the European exchange rate mechanism, inflation was relatively low during the last decade. As inflation was low and real interest was relatively high, the consumers had a stimulus to invest in long-term oriented savings because there was no fear for negative returns and the average long-term interest rate offered by insurers (actuarial interest rate) was more competitive than in a high inflationary environment.

We already mentioned that also saving trends have changed. Hoschka (1994) describes it as follows:

> "In particular, life insurance accounts for an increasing share of total savings in some EC countries which underlines its rising significance and the successful inroads which insurers were able to make into the savings pool which, traditionally, was virtually monopolised by banks" (p. 30).

Here we present some figures of the German situation in the 1980s. (For more details and other statistics see chapter 2.)

Thus banks saw an outflow of their deposit base towards life insurance companies. It is no surprise banks have tried to recapture their cheap ways of funding by diversifying primarily into the life insurance industry.

Figure 1.1 Development of total financial assets in private households and share of insurance and pensions (Schwebler, 1989)

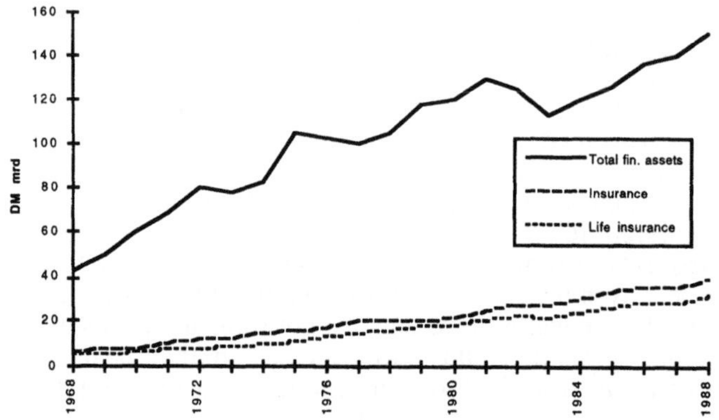

This strategic move of banks into insurance was further stimulated by other factors such as the growth perspective in the insurance industry, the support of an advantageous tax regime, and the gradual regulatory liberalisation within the financial services industry. The regulators realised it was not necessary to rely on direct interference by imposing restrictions on activities of banks and insurance companies. In the Netherlands, the liberalisation of the *Structure Policy* in 1990 has been one of the main causes of the accelerating pace of take-overs in the financial services sector.

Different types of diversification strategies are possible: methods of entry

A wide range of organisational structures has been developed to form these financial conglomerates (Lafferty Business Research, 1991; Koguchi, 1993; Hoschka, 1994). Some of these groups opted for rather tight relationships and merged with, acquired or set up an insurer or bank. Others have chosen looser forms of co-operation, such as joint ventures and marketing agreements.

According to Hoschka (1994) and Lafferty Business Research (1991) the most successful strategy is *starting from scratch.* A financial institution starts a separate company to sell non-core products, e.g. an insurance company starts up a bank or a saving institution; a bank starts up an insurance brokerage and/or an insurance company. In this way, the business and marketing style can be designed straightforward, without entanglements of an existing company, diverging opinions and conflicting cultures or images.

However, it will take a longer period before a good level of activity will be reached. Furthermore, the company can make no appeal on a partner so that technical expertise and know how will have to be brought in from outside sources.

A second strategy is to *acquire* or to *merge* with an existing company. The differential advantage of this type of diversification mainly lies in the fact that the expertise and experience necessary is available from the start. One of the main criteria in the search process of potential alliance partners must therefore be the knowledge and professional skills available. The appropriate types of skills and knowledge will of course be totally different according to the product and market segments the company wants to tackle. A mismatch here can be an expensive misinvestment (Van den Berghe, 1989).

This strategy has a number of disadvantages. Besides the problem of finding a good partner, mergers and acquisitions are also very expensive. Diverging cultures and management styles can also cause a lot of distortions.

These problems can differ according to the type of merger or acquisition. Mergers and acquisitions indeed cover a rather heterogeneous set of structures and strategies. Some have far-reaching goals of integration and co-ordination so that one global strategy and structure is installed. While others are merely focusing on the diversification of a portfolio of businesses or aiming at a defence against competitive pressures, leaving the merged activities more or less intact. In case of an acquisition the strategy can be enforced by the parent company who has the power to outline the business policies. In case of a merger, it will probably depend on the aim of the merger and on the respective weight of both partners -the bank and the insurance company- which direction (e.g. bancassurance and/or assurfinance) is finally chosen.

Another method of entry which is frequently used, is the *joint venture*. A joint venture is considered to be a more formalised way of co-operation. However, different degrees of co-operation can be observed (Stafford, 1994). A creative joint venture is defined as an agreement between two or more independent companies, setting up a legally independent entity owned and controlled by the parent companies. An acquisitive joint venture is based on a equity participation or cross-shareholding to foster collaboration in some of the business activities. Joint ventures can be used for several reasons:

- a hand-over strategy only looks to outsource certain business tasks to the joint venture partner;
- a trade strategy wants to hedge on each other's distinctive competencies;
- a pool strategy goes further in the pooling of resources and know how.

"As alliances move from hand-over strategies to trade and pool strategies, integrated value chain activities become progressively more intricate with increasing partner interdependency and co-operative inflexibility" (Stafford, 1994).

Most of the joint ventures between banks and insurance companies concern life insurance companies : the insurance company brings in the know how and the expertise, the credit institution provides the distribution facilities.

Nowadays, joint ventures are subject to a lot of research, which has yielded remarkable results. It is argued that with this diversification strategy, one focuses too much on the short-term profitability thereby neglecting effects on the longer term. Bleeke and Ernst (1995) have found that "the median life span for alliances is only about seven years, and nearly 80 per cent of joint ventures -one of the most common alliance structures- ultimately end in a sale by one of the partners".

Nevertheless, joint ventures are frequently used especially with cross-border alliances. In this way, banks and insurance companies from more mature markets tie with local insurers or banks. By doing so, they avoid problems which could result from a lack of knowledge of the market situation. Furthermore, risks (but also profits) can (have to) be shared with more partners.

To cross-sell other (than core business) products sometimes *marketing agreements* can suffice. They are used when firms want to test the water before plunging into more expensive forms of co-operation. This strategy gives firms a high degree of freedom, but marketing agreements are often completed (after a while) with (substantial) participations of the partners into each other's capital, hereby entering the joint venture route. This is often a sign that the established relation between the two companies gets a lasting character.

Marketing agreements can also refer to other aspects than distribution issues, but in most cases it started with distribution agreements: banks distribute insurance products through bank branches while the insurance companies sell savings, investment and/or credit products through their sales channels.

As we already mentioned, this is quite a loose form of co-operation compared to the other approaches and it does not always lead to the expected results. The combined offering of insurance and financial services requires considerable involvement to resolve the bottlenecks that will surely appear. With this strategy, however, it can be feared that the two parties will give up efforts more quickly and will pay attention to their own priorities. The French bank, Banque Nationale de Paris (BNP), is a good example in this respect. When they saw the distribution agreement with UAP did not yield the expected results, they focused on a co-operation agreement with the German Dresdner Bank.

2

FROM BANCASSURANCE AND ASSURFINANCE TOWARDS ALL FINANCE - THE CONCEPTS OF UNBUNDLING AND REBUNDLING

HEADING TOWARDS BANCASSURANCE AND ASSURFINANCE

In the first chapter we have given some arguments why banks have entered the (life) insurance industry. The growing pressure on their profit margins made them look for other sources of profit which they found by acting as a broker for insurance companies. This process has been going on for years and banks had reasonable success by selling insurance products through the bank branches. It is interesting to notice that banks also have diversified into other directions, e.g. investments were made in the real estate industry. The severe problems, some of the French banks experienced, show that this move was not always without risks.

While many banks were glad with the existing relations, others were thinking of operating the insurance business themselves, because they felt they were losing the battle in the savings market. By becoming insurers themselves, they could attract some new types of insurance savings, which enlarged their deposit volume with stable contractual savings. Moreover, these types of insurance savings seem to be growing at a much greater speed than any other savings market, probably thanks to the growing awareness of the need for individual responsibility in the field of old age pensions.

The figure below gives an overview of the situation in the Netherlands from 1955-1993. A distinction is made between contractual savings (by life insurers and pension funds) and free savings (all other forms of savings).

Looking at Figure 2.1, we can conclude that contractual savings have grown much faster than free savings. This growth has mainly been caused by the success of pension funds. It is only from the middle of the 1980s that life insurers also have contributed to the growth of the contractual savings.

Figure 2.1 Free and contractual household savings

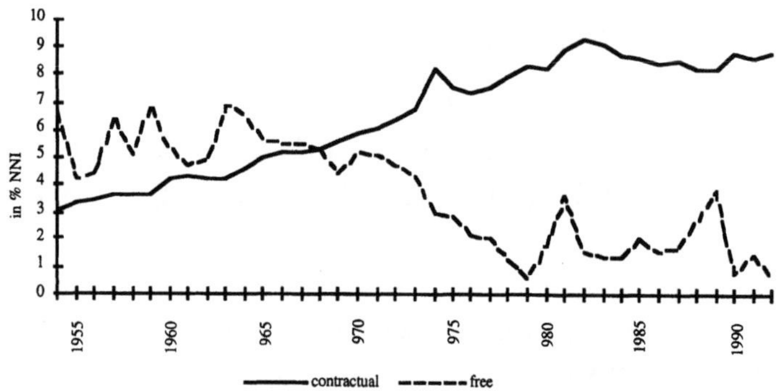

This trend can also be seen in other countries. The following table gives an insight in the evolution of the outstanding financial assets in the household sector of different countries.

Table 2.1 Outstanding financial assets in the household sector (shares in total assets; per cent)

Canada	**1976**	**1981**	**1986**	**1991**
Cash & Deposits	38	39	34	36
Bonds	11	10	10	9
Shares	23	24	23	20
Insurance & Pension	20	22	26	29
Others	9	6	7	6
Total	100	100	100	100
(Total: C$ billion)	(272)	(529)	(837)	(1 246)

France	**1977**	**1981**	**1986**	**1991**
Cash	20	20	14	11
Deposits	45	42	30	25
Bonds	8	9	7	4
Shares	9	13	35	42
Insurance & Pension	3	3	5	10
Others	14	13	10	8
Total	100	100	100	100
(Total: FF billion)	(1 900)	(3 246)	(6 767)	(10 129)

Table 2.1 *(continued)*

Germany	1976	1981	1986	1991
Cash	9	8	7	8
Deposits	53	50	45	39
Bonds	11	14	14	18
Shares	6	4	7	6
Insurance & Pension	19	22	24	26
Others	2	2	3	3
Total	100	100	100	100
(Total: DM billion)	(1 041)	(1 598)	(2 338)	(3 203)

Japan	1976	1981	1986	1991
Cash	17	13	10	9
Deposits	58	61	52	52
Bonds	9	10	10	8
Shares	3	2	11	8
Insurance & Pension	13	14	16	21
Others	0	1	1	1
Total	100	100	100	100
(Total: Yen trillion)	(196)	(368)	(650)	(992)

Spain		1982	1986	1991
Cash		14	13	18
Deposits		52	49	42
Bonds		5	10	6
Shares		14	14	16
Insurance & Pension		2	3	8
Others		14	11	9
Total		100	100	100
(Total: Ptas billion)		(21 956)	(37 477)	(64 059)

United States	1976	1981	1986	1991
Cash	4	4	4	4
Deposits	22	21	21	18
Bonds	7	7	9	11
Shares	19	16	20	20
Insurance & Pension	17	17	24	28
Others	31	35	23	19
Total	100	100	100	100
(Total: US$ billion)	(3 992)	(7 039)	(10 804)	(15 245)

Source: Koguchi (1993)

The fiscal treatment of insurance savings still seems to be a factor that motivates people to invest their money in insurance products. We have already pointed out that the outflow of banking deposits made the banks decide to set up their own insurance companies.

The emergence of financial conglomerates has not only to do with factors concerning the macro-economic environment, but also with the successes of some early moves. The pioneers operated in France and Great Britain. One of the most interesting success stories was Crédit Agricole which has set up Prédica, a life insurance subsidiary, in 1985. This company delivered simple and transparent life insurance products *-bons de capitalisation-* through the branches of the bank. The success was enormous and after only seven years of operation Prédica had become the fourth largest life insurer in France. Part of the success however must be contributed to the fiscal advantages of the product. In Great Britain the insurance subsidiary of former TSB and Lloyds Abbey Life, the insurer of the Lloyds Bank, were the success stories of that moment. In Germany Deutsche Bank Leben showed remarkable growth but after a few years was incorporated into the Deutscher Herold Holding, an insurer linked with the Deutsche Bank.

Success breeds success and soon the pioneers were followed by most of the larger financial institutions. At the other hand insurance companies who faced an invasion in their territory, decided to step into the financial services industry, either by starting a bank from scratch or by forming distribution agreements. There were also quite a number of mergers between banks and insurance companies, establishing tight relations between the two different parts. As we have already mentioned: the interface between banks and insurance companies became competitive, and from this moment on the first articles on *bancassurance* (and *allfinanz*[1]) appeared although banks had been selling insurance products through their branches for years.

In financially less developed countries, foreign banks formed joint ventures with local insurance companies or vice versa. In 1990, TSB Trust Company (GB), Caisse Nationale de Prévoyance (F) and Cariplo (I) set up Carivita, a life insurance subsidiary which has marketed life products through Cariplo's branches in Italy. After its success in the French home market, Prédica, the life subsidiary of Crédit Agricole, formed joint ventures in Italy (with Monte dei Paschi di Siena) and in Portugal (with the Espirito Santo Group).

After all, it is not surprising to realise that banks and saving institutions took the lead in this diversification process and in the financial services revolution.

- "... The scope for rationalisation resulting from the new delivery systems is so great, that all retail banks will have to find new ways of making better use of their existing branch network. This is likely to mean further attempts to develop their role of financial advisers as opposed to simple money processors" (A.D. Little, 1985).

---

[1] A lot of these early studies used the terms *bancassurance* and *allfinanz* (*all finance*) as synonyms. However they did not make the distinction we made in the first chapter.

- The introduction of large-scale automation also in the relations with the customers (e.g. Automatic Teller Machines) creates a need for more high touch products.

IS IT TRUE THAT BANCASSURANCE IS MORE SUCCESSFUL THAN ASSURFINANCE?

Bancassurance versus assurfinance

Although it is certainly preliminary to draw hard conclusions from the recent trends towards bancassurance, assurfinance and all finance, the initial signals prove that the sceptics could be completely wrong in disapproving this trend. Moreover it is clear that the success of some initiatives in this respect has attracted greater attention for and interest in this diversification. The *demonstration effect* has indeed resulted in a far greater number of diversified firms, whereby nearly all market leaders in the different European countries are involved in this process. The synergies seem to be straightforward because of the rather close relatedness between both markets. Nevertheless the success of bancassurance was first considered to be very unsure which is described perfectly by this statement:

> "The prospect of banks selling insurance is a little like Beaujolais Nouveau's arrival, each year: every one gets excited, but no one knows whether it is any good" (Lloyd's List, 21 Nov., 1988).

Detailed and historical databases on both bancassurance and assurfinance trends are not available (yet). Nevertheless, there are indications that the bancassurance trend has been the more successful one. Lafferty Business Research (1994, p. XVII & p. 249-250) mentioned a study by Consilium (a German strategic management consultancy firm) which evaluated about 20 European Allfinanz initiatives (results are shown in figure 2.2).

> "As a marketing initiative, Allfinanz has been very successful for some institutions... The analysis shows, as often stated, that only the large retail banks stand to gain from an Allfinanz initiative."

Banks have the advantage that they possess the primary current account which Consilium considers to be the main cross-selling platform. The current account is the source of customer information and is the basis on which banks can provide superior service.

> "The permanent access provided by the current account, together with the ability to identify the right sales opportunities, allows the banks to manage their customer bases and maximise the cross-selling potential" (Lafferty Business Research, 1994: 250).

A limitation of this strategy is that it is limited to the middle income segment because only this segment generates sufficient extra business at a reasonable price. On the other hand, extra costs of training and M.I.S. must be incurred and this necessitates a certain scale (which is also demonstrated by figure 2.2).

Figure 2.2 Consilium ranking of European all finance initiatives (Lafferty Business Research, 1994)

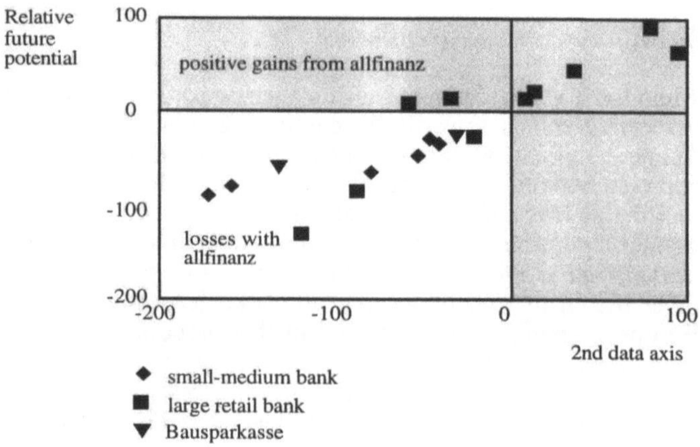

Other data sources have come up with similar findings and have argued that banks, compared to insurance companies, are in a better position to implement a diversification strategy. In general the better performance of bancassurance is explained by the following factors:

- economies of scope can be realised better and quicker within the banking firm;
- banks have a far better distribution position, thanks to the banking account.

Banks have been successful in competing with traditional insurance intermediaries due to some competitive advantages:

- They had an extensive branch network at their disposal which could be used to sell insurance products.
- They had more frequent contact with their clients. This advantage is decreasing rapidly due to technology-driven banking services, and insurance selling can therefore be seen as a hedge against this leakage of client contact.
- They could use a much more detailed database to approach the customer and deliver tailor-made products with a one-stop-shopping convenience.

- For the most part they had a better reputation: banks were seen as *trust persons*, as opposed to insurers who were thought of as *sellers*.

The moves of insurers into the retail banking business were less noticed and did not always seem very successful. In Germany, the rapidly expanding insurance group Aachener und Münchener ran into difficulties when they bought a former trade union bank, Bank für Gemeinwirtschaft (BfG). The problems had become too serious and therefore, they reduced their participation in the bank to 25 per cent. Other problems originated in France where GAN, one of the largest insurers, had problems with its subsidiary CIC, because the latter one was too heavily exposed to the collapse in the French property market. Similar problems occurred in Australia (Chase-Australian Mutual Provident's) and New Zealand (General Accident and NZI Bank).

It is remarkable that insurance companies played a less important role as agents for banking products. Possible explanations are:

- regulatory restrictions: banks could sell insurance products without great regulatory barriers, whereas insurance companies could not easily attract public savings (deposits, term accounts, etc.) or operate as a monetary institution;
- their institutional investor function made insurers important players in the investment field as well as in the field of mortgage loans;
- distribution facilities: most banks (except the savings banks) operate through their own distribution network. This gives them the opportunity to implement whatever marketing strategy they want. Insurers on the contrary operate in many countries through independent brokers so that they do not reach their final clients directly.

In the past, independent intermediaries themselves have, to a certain extent, been involved in the distribution of financial services: they not only offered mortgage loans (coupled with life insurance and eventually with fire insurance) but many of them operated also as agents for a savings bank.

Research in Belgium (Van den Berghe and Baeten, 1994-1995) has shown that 65 per cent of the independent small and medium-sized intermediaries also offered financial services. However, the contribution of these financial products towards the turnover is much smaller: financial products represented only 17 per cent of the total commission earned by the interviewed intermediaries. The market percentage of profits attributable to the financial products can even be lower because it is frequently heard that the commissions on these products are (very) small compared to those of insurance products.

But there are potential threats as well

One important aim of most diversification strategies of financial firms is the realisation of a multi-channel approach. Since most diversification strategies aim at or start with the realisation of the benefits of cross-selling, this aspect is relevant for almost all diversifying players, regardless of the possible differences between them in implementing their strategy.

By far the most important barrier in relation to cross-selling exists for insurers operating through a network of independent intermediaries. Most brokers consider the multi-channel approach as an attack to their position. They also fight against the so-called dual pricing, whereby a diversified firm adopts a different price for its products according to the type of distribution channel used. Also the potential use of client data bases for cross-selling is under attack by intermediaries who proclaim that insurers are not allowed to contact *their* clients directly (without explicit agreement of the respective intermediary) or transfer client data bases to other business units (which do not operate through brokers).

Example 1: the case BNP-UAP

These problems were faced by UAP (Union des Assurances de Paris) which was the largest insurance group in France before the merger with AXA. The original purpose of UAP was to sell non-life insurance products through the bank branches of BNP (Banque Nationale de Paris). Conversely the agents of the French insurer would sell home mortgages, car loans and credit cards of BNP to their clients. But while the relationship between the two institutions was becoming closer, UAP was put under severe pressure from its main distribution network consisting of 1600 exclusive agents who were disgruntled about the increasing importance of BNP outlets as alternative distribution channel for UAP (Lafferty Business Research, 1991). The agents refused to sell the products of BNP and when the insurance company offered them the possibility to step out of the contractual obligations, many have accepted this proposal and left. In this way they have expressed their disagreement about the co-operation with BNP.

Example 2: the case ING

A similar event occurred in the Netherlands when ING was formed in 1990. ING was the result of a merger between Nationale-Nederlanden (NN), the largest Dutch insurer, with the banking group NMB-Postbank Groep. NN mainly used independent brokers and at that time they accounted for more than 80 per cent of total premium income. However when NN decided to develop insurance products for the Postbank, a direct bank with 6 million customers, the independent brokers boycotted the sale of NN policies. Only when the management of NN announced that the Postbank would be used for the distribution of simple standard products, the relations with the brokers could be restored.

There are other threats as well. Insurance marketing differs to a considerable degree from the marketing of banking products. Banks have a (far) better image, not in the least because of passive servicing attitudes and a better pull strategy. Insurance companies on the contrary have a less favorable public image and this in relation to the aggressive selling methods often used in practice. Moreover their reliance on a push strategy and the negative connotations of risks hamper their possibilities to create a better company image. If banks become underwriters, their image could be negatively affected because of the tough responsibilities of insurance claims handling. Basecq, CEO of ASLK Verzekeringen (an insurance subsidiary of Fortis), compares this danger to a *bombe à retardement* (Feiner, 1995). However, in practice the gap between the respective push and pull strategies seems to be decreasing.

Also the remuneration structure used is quite different for insurance products (mainly commission as a percentage of turnover) and banking products (mostly fixed income complemented with a variable bonus that relates to quite a large number of qualitative and quantitative performance measures). Consequently an insurance intermediary used to sales incentives is not always interested in the remuneration methods of banks, whereas bank employees sometimes feel less motivated to make extra efforts to sell insurance products without extra remuneration stimuli.

Another aspect that deserves our special attention is the following: research shows that customers seem to have quite distinct ideas about this whole diversification trend:

- a considerable number of consumers claim to be interested in the so called one-stop-shopping, offered by diversified financial firms;
- on the other hand managers experience problems with customers who do not agree to buy different types of financial services from one kind of provider.

Research we conducted in the Belgian market over the period 1988-89 and 1992-93 revealed that in the beginning (1988-1989), the majority of the Belgian population did not support the one-stop-shopping idea. This originated from a feeling of distrust or a lack of control. Only one out of the three clusters found it had advantages to buy financial services of one provider. This cluster, which represented a small 40 per cent of the Belgian population, was rather passive and trusted the financial institutions. Their attitude could be explained from a convenience point of view. However, these differences in attitudes did not lead to differences in purchasing behaviour, partly due to the lack of a diversified market supply.

The results from the market research in 1992-1993 made it clear that the attitude towards one-stop-shopping had become more positive. It should be noticed that the qualitative research revealed that the image of the banks had become less positive; this is partly due to the harder selling techniques, recently introduced in the bank offices; another explanation could be found in the discontent of consumers with the tarification of banking services (previously they were free of charge).

Only one out of the five clusters, representing a fifth of the population, had a real problem with integrated financial services. This attitude could be explained by the same reasons as found in the market research of 1988-1989: a feeling of distrust and a lack of control.

The *trustful people* (18,3 % of the population) were proponents of an all finance approach. To them both the bancassurance form as well as with the assurfinance form offer a lot of advantages. According to the other clusters there was a difference between assurfinance and bancassurance. Generally, the latter was preferred. These respondents did not feel very comfortable with the idea of insurance intermediaries who also offered other financial products. However, the cluster of the *nostalgic risk avoider* preferred a relation with one contact person. They chose the insurance intermediary because they felt more at ease with persons of their own 'level'. This market research not only revealed differences in attitude but it showed that this difference also led to a diversified market behaviour. The distribution of financial products is still less diversified than the channels through which insurance is purchased, thereby proving that bancassurance is further developed than assurfinance.

Are the synergies between banks and insurance companies limited to the retail market, and mainly to the life insurance market?

It is often stated that banks preferred life over non-life insurance. Hoschka (1994) analysed 35 EU banks which engaged in bancassurance and discovered that more than 80 per cent had only entered life insurance (figures of 1991). 5 per cent of the banks have entered non-life insurance, while the remaining 15 per cent was active in both markets.

We can also discover this trend by looking at the penetration of banks in the insurance market. In 1993, Belgian banks had a market share of 14.25 per cent in the life market, while only 3.11 per cent in the non-life market. More recent figures are provided in table 2.2. The table shows the premium income of the largest bancassurers in Belgium and indicates how much premium income was earned through the bank branches in both 1995 and 1996. All companies listed are insurance companies who distribute their products solely through the bank branches. We can see that the banks have achieved considerable growth in their total premium income. However, the market share percentages remain relatively constant, which indicates that the traditional distribution channels for insurance products have also performed well.

Table 2.2 Bancassurance in Belgium: the results in 1995 and 1996

Bancassurer		**1996**			**1995**	
	Non-life	**Life**	**Total**	**Non-life**	**Life**	**Total**
ASLK/CGER	4.539.227	26.545.587	31.084.814	4.130.318	25.620.603	29.750.921
Omniver	1.905.532	5.836.755	7.742..287	1.489.195	6.249.849	7.739.044
Alpha Life	-	7.060.164	7.060.164	-	6.705.367	6.705.367
Mega	1.513.709	4.835.137	6.348.846	2.092.488	2.127.710	4.220.198
BBL Verzekeringen	86.815	4.106.728	4.193.543	77.403	2.499.030	2.576.433
Aras	1.432.708	1.245.021	2.677.729	1.233.357	207.275	1.440.632
Citilife	-	961.907	961.907	-	923.183	923.183
S-Leven	-	550.900	550.900	-	369.742	369.742
HBK-Leven	-	159.282	159.282	-	-	-
Eural-Unitas	-	34.947	34.947	-	33.374	33.374
Total Top 10	9.477.991	51.336.428	60.814.419	9.022.761	44.736.133	53.758.894
Total insurance market	333.013.061	216.905.165	549.918.226	319.847.118	187.164.952	507.012.070
Market share Top 10	**2.8%**	**23.6%**	**11.0%**	**2.8%**	**23.9%**	**10.6%**

Source: De Financieel-Economische Tijd (27 December, 1997)

The dominance of life insurance in the bancassurance portfolio is remarkable. This is a historical phenomenon because banks have always been more interested in selling life insurance products. And the establishment of life insurance companies in the 1980s and the beginning of the 1990s must been seen in this context. This can be explained by the fact that the banks were interested in an enlargement of their deposit volume with stable contractual savings. As we have shown before, banks have been confronted with a smaller deposit base as a consequence of an outflow of savings to other financial institutions. Because premium income of life insurance exhibit a much more stable character (thanks to the long-term character of the business), also banks were interested in this kind of business. Furthermore, the activities of life insurance companies had more in common with those of banks and other credit institutions. General insurance was seen as a kind of business which was totally different from the bank's core business.

However, some leaders in the all finance movement proclaim that the actual focus on the retail market co-operation between banks and insurance companies is only the *tip of the iceberg*. Also the corporate market could offer many opportunities for joint and integrated projects. And in chapter 8 we will discuss the growing links between investment bankers and reinsurance companies.

ALL FINANCE AND (RE)BUNDLING OF SERVICES: FROM FACT FINDING TOWARDS CONCEPTUALISATION

In Chapter 1 we already discussed that a distinction has to be made between bancassurance and assurfinance on the one hand and all finance on the other hand. This distinction is relevant from many perspectives:

- it helps to analyse the many variations in the diversification strategies of banks, insurance companies and other financial institutions;
- from a management perspective this clarifies the many diverging implications on the level of marketing and strategy as well as on the organisation and the *production process*;
- this distinction is necessary from a more theoretical point of view;
- moreover the in-depth analysis of this distinction can serve as a valuable yardstick to upgrade the general literature on the bundling of services with some valid case-material and practical experience.

How can the (re)bundling of services be defined?

The traditional definitions of most financial and insurance services are based on technical criteria and institutional sectoral barriers. The more bankers and insurers observe the real needs and wants of their clients, the more they realise that these traditional product concepts are often not in accordance with the real market needs.

The complementarity in time and space between different financial and insurance products creates natural incentives for cross-selling, packaging and product integration. Through unbundling (old traditional products) and rebundling (in accordance with the real market needs) new service bundles are created.

Bundling can mean different things to different suppliers or customers. It is therefore necessary to analyse into more detail the elements that differentiate the many forms of bundling that exist in practice. This analysis will help to make the distinction between bancassurance and all finance.

The many forms of bundled services can be differentiated with the help of the following criteria:

- the level of customisation and integration between the different products and services involved;
- the underlying degree of expertise with the service provider for each of the products and services involved;
- the value added for the customer, which finally is the ultimate goal of this bundling.

These three elements can be combined in one graph to give a visual picture of the types of bundling.

Two extreme cases can be observed and in between, a whole bunch of variations are possible.

- At the one hand we have *cross-selling* with the lowest degree of integration and customisation, with a limited scope of different expertises involved and a relatively lower level of value added. Here the accent of the supplier's strategy is more on the volume of business and this from a perspective of transactional marketing.
- At the other end we find the *fully integrated services* with a high degree of integration and customisation, with many experts involved and a high value added for the customer. Here the supplier is looking for a competitive edge in tailoring different complementary products and services, personalised from a lifetime perspective or built around specific needs or events. This is a typical approach for relationship marketing and builds on customer loyalty.

Figure 2.3 The different forms of bundling

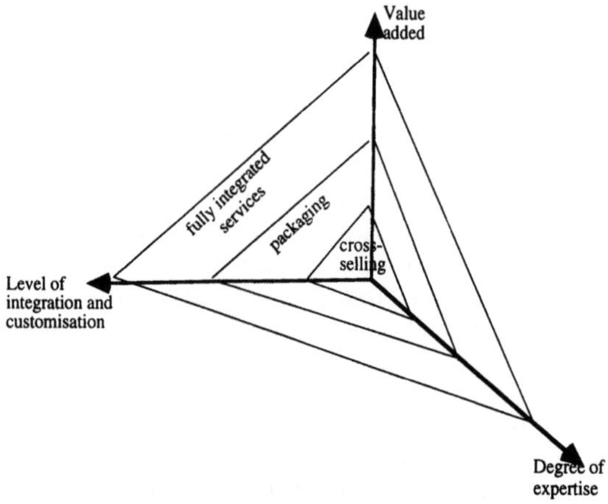

- In between all types of *packaging* can be found with intermediate positions on the level of integration, customisation, expertise and value added.

This evaluation is based on the approach developed by J. Heskett e.a. (1997). The value added for the client depends upon the following elements:

$$\frac{\text{result (3) + process quality (4)}}{\text{price (1) + non-monetary costs (2)}}$$

The definition of Heskett e.a. is very useful to evaluate the extra value added that can potentially be realised with integrated service bundles and shows where possible negative side effects can prevail.

(1) Thanks to economies of scale as well as economies of scope the *price* of bundled services can be lower than that for the services bought separately. In practice, most bundled services are sold at a price discount. It can however be more difficult to compare the respective prices because bundled services are often less comparable than the standard ones. Critical customers can evaluate this lower transparency as a negative side-effect of bundling.

(2) The *non-monetary costs* consists of different subelements:

- The *search costs* will decrease drastically if a supplier offers bundled services. Especially less-informed customers can find it very difficult and they are even afraid to shop around and to look for the most convenient supply in the market. This is even the case for more informed customers when it comes to complicated products and complex determinants of the best needs. Quite a number of financial products are rather complex.

- On the contrary, the *switch costs* or the cost of changing suppliers will increase as the integration of the services involved grows. Moreover suppliers often expect that customers will sign more longer-term contracts when dealing with customised or personalised integrated service bundles. These factors can be an argument in favour of the supply of bundled services: the suppliers buy more loyalty because of the longer-term contracts and the higher switch costs for the customers. However they can entail a disadvantage, especially for critical customers and shoppers. These observations are very useful wen applying these theoretical concepts in practice: our market research (Van den Berghe and Baeten, 1994-1995) clearly showed that some customers are considering this non-monetary cost as the main hurdle to accept the one-stop shopping idea of bundled services or the integrated approach of all-finance. Luckily for the suppliers not all customers feel the same way: some customers do not find these non-monetary costs that important because they attach more value to the higher quality and result. Our research also revealed that even critical customers can be convinced of the net surplus-value of integrated services if they see that the advantage on the price level is more important than the disadvantage on the level of the switch cost.
- Another element of the non-monetary costs is related to the fact that bundling leads more or less to the principle of *all eggs in one basket*; especially critical customers can see the different potential disadvantages in this respect. They fear that one institution will know too much about them: this is a real fear in relation to financial services. Some also see a potential danger of *concentration of power* within the hands of the service provider. These dangers are often expressed by well-informed or critical customers, whereas less eloquent customers often feel more comfortable when they do not have to give details about their personal needs, situation and problems to every service provider. A familiar intermediary is often best situated to become the confidant for the personal financial diagnosis and recipes.

(3) The more the services will be bundled and integrated the higher the *result* for the customer and this from different perspectives.

- Suppliers of integrated services often offer a *result commitment*, thereby working more solution-driven than volume- or product-driven. In the case of a pure cross-selling approach it is very well possible that over-selling goes hand in hand with white spots or failing covers of service components. Eliminating these over- and undersupply through integration and customisation creates a higher value added for the customer and a better competitive position for the supplier.

- The integration also supposes a considerable degree of individualisation or at least customisation. This certainly improves the value for the customer who receives solutions fine-tuned to his specific needs and expectations.
- The components of the bundled services can be produced by different *specialists* as far as this is done by a large conglomerate or an integrated network of specialist service providers. According to many management experts, the networking road looks promising, but there still remain some dangers. Firstly, are all partners in the network delivering the same quality? The chain is as strong as its weakest link: therefore bundled services require an integrated management of the whole service supply-chain. Secondly, how strong is the network? Will the collaboration last? This question is especially relevant in sectors like insurance, where a customer buys a promise that will have to be honoured in the future, which can sometimes be 30 to 40 years away from the signing of the contract (e.g. long-term care contracts or pension provisions).

(4) The *process quality* will be improved from different perspectives:

- The one-stop shopping is more convenient.
- The customisation supposes a relationship marketing whereby the client receives a personalised approach instead of a standard client approach of the transactional marketing type inherent in cross-selling.

What factors are the driving forces behind the trend towards bundling of services?

It is not only the financial sector that is witnessing a trend towards bundling of services. In other service sectors as well as in industrial activities, bundling of services is becoming an important strategic and marketing tool. There is a lot to be learned from analysing and comparing the drivers behind all these strategies.

The *market environment* is a very important driving force towards the bundling of services. This is certainly the case where a tough competitive struggle is taking place, e.g. in the financial and insurance sector. Hectic competition leads to more price competition and creates a danger of a negative price spiral and even a price war. In such a difficult market environment the suppliers are intensively searching for an alternative competitive positioning and new competitive advantages. The bundling of services is seen as a method to improve the competitive position both from a price/cost and a value perspective. The deregulation and privatisation of part of the social security system can pave the way for more diversification and bundling of insurance and saving products. Especially the Dutch market is an interesting test market to observe. Most major players in the insurance sector that formed bank-insurance groups have recently joined forces in networks with organisations for social security administration. For example, the ING Group teamed up with the SFB Group while Achmea joined GAK and the Rabobank-Interpolis Group formed an alliance with GUO-Cadans.

The success stories in all finance have led to many new ventures. Illustrative in this respect is a passage from an interview with Lukas Mühlemann, a former McKinsey management consultant and now chief executive officer of the Crédit Suisse Group.

Interview with Lukas Mühlemann (FT, 31 October, 1997)

Question: What will the European banking landscape look like in five years?
I see two forces at work: the on-going shake-out within the different financial markets and the coming European-wide consolidation with regard to the introduction of the euro. National markets will become less and less important as currency borders disappear. In this sense the euro can be regarded as a catalyst. As a result the market will be carved up among the few truly financial global players, even though there will be room for smaller banks to focus on niches.

Question: The recent spate of mega-mergers in the US financial services industry is spilling over into Europe. What is the driving force?
The liberalisation of the US financial market (introduction of inter-state banking and the relaxation of Glass-Steagall Act) have increasingly led to consolidation in commercial and investment banking. In Europe, we will see a similar development, although for different reasons. European banks, which have already entered the bancassurance era, are looking for size, whereas the US banks want to get ready for the era after Glass-Steagall.

Question: Crédit Suisse has set its sights on becoming a leading bancassurance company. Why do other banks not share your enthusiasm for the concept?
The bancassurance sceptics are increasingly losing ground. As we have seen since we announced the merger with Winterthur, others are following the example set by ING, Lloyds and Crédit Suisse.

Also the expectations of the customers are forcing the suppliers more into the direction of integrated services. The abundant supply gives them the choice to shop around both for a good price and a good service at the same time. Furthermore they no longer accept standardised products that do not fit their specific needs but look for tailored solutions. This is no longer only relevant for the industrial clients or the up-scale market but is increasingly the case in the middle market too. The overwhelming variety of products and services makes it more than ever necessary that the customers are guided through the information overload of choices and customisation questionnaires.

These observations are especially valid for the financial sector. Our market research in the financial and insurance sector in Belgium and the Netherlands (Van den Berghe and Verweire, 1992-1993; Van den Berghe and Baeten, 1994-1995) clearly showed that most customers are not at all happy with the fragmented product approach of insurance companies. They do not want a set of insurance policies, whereby they feel uncertain whether they are insured once a risk happens, but they are in great need for customer-friendly solutions. In the past insurers have been focusing on the technical risk categorisation and tarification criteria at the one hand and on the legal and contractual fine-tuning at the other hand, thereby using a very supply-oriented approach and an inaccessible jargon in their product development. Their customers expect however they come up with solutions that preferably take into account a bundle of risks or a bundle of needs in relation to a certain event. Their customers don't want to be bothered with the ingredients and the recipe, but are looking for the prepared dish.

What factors enable companies to enter into a bundling process?

A number of elements enable the development of bundled services. Some of them are even so important that they form the conditio sine qua non for a successful bundling strategy. Modern information technology (IT) is certainly one of these elements for several reasons.

- IT is a precondition to set up a customer-oriented database and to develop a pro-active approach towards one's customers' needs and expectations.
- Secondly IT is the most crucial factor to go from customisation and individualisation towards mass-customisation and mass-individualisation so that customers can allow themselves to pay for these services.
- The networking between different producers and service providers is only feasible when they are linked through high-tech networks.

As we mentioned before, the different producers and service providers have to *collaborate* in order to arrive at a bundling of services. To achieve this a number of firms have to *create networks and form alliances*. In this respect we could refer to diversification strategies but bundling of services is more than diversification. It often goes together with some form of specialisation whereby one focuses on the core business, but also involves outsourcing as well as networking. The network can be built around the management of the value and the supply chain. Whereas customisation only integrates the supply chain at the end phase of the distribution, individualisation entails that the whole value chain is re-organised to meet the customer's specific needs and expectations. This *rethinking of the service chain* has far-reaching consequences for the need of an integrated management, administration and organisation of all the suppliers involved in the service bundle: the chain will only be as strong as it weakest link.

This factor is of the utmost importance for service providers that operate with the help of independent intermediaries (as is the case with many insurance companies that operate through a network of brokers) who want to shop around for every product within the bundle offered. Only a modern definition of *independence* can help to overcome this huge barrier: the modern philosophy of networking within a democratic corporation supposes a mature attitude of openness, thrust and long(er)-term commitment. This also supposes the right philosophy, attitude and experience with respect to modern networking. These new trends have been analysed very well by Ackoff in his book on the *Democratic corporation* (1994).

The choice for a *relationship marketing* approach instead of a transactional marketing option is another important determinant in the bundling strategy. In the beginning such a choice can be much more expensive because it needs more up-front investments. But in the longer run such an approach offers quite substantial advantages: the long-term relationship with the client leads to a learning relation that is upgraded with every client contact. The loyalty of the customer is honoured with ever more personalised service at more or less a *mass* price. Consequently, successful firms can turn the vicious circle of price discounts caused by cross-selling into a virtuous circle of a loyal client relationship.

Some examples

Many types of bundled services are developed at great speed in the market for financial services. Although it is not always easy for an outside observer to evaluate the degree of integration and customisation as well as the level of expertise of the service providers and consequently the value added for the customer, some broad lines are clearly observable.

Typical examples of *cross-selling* are the supply of life insurance products by an insurance intermediary, specialised in non-life products such as car insurance and family liability covers. Another example in the banking sector is the offering of saving accounts as well as current income accounts. Diversification of financial firms and the blurring of the boundaries between the different sectors enlarge the scope of cross-selling to an assortment that contains insurance products as well as banking and other financial services.

The insurance sector as well as the banking sector increasingly realise that customers are better served if they offer them a more fine-tuned offer of *packaged services*. Good examples in this respect are for instance the unit-linked products that combine savings, investment and insurance components. Also the packaging of mortgage loans with a number of insurance covers (death, fire) or a car loan with a car insurance is common practice.

Packaged insurance products

In a Dutch magazine intended for the insurance intermediaries an overview was given of personal insurance packages. Besides the name of the insurer and the insurance package we also have mentioned the year of introduction and the percentage of the new production coming from these packages.

Aegon	1995	Woonpakket-plus	60%
De Amersfoortse	1994	Domino Pakket Plus	
AMEV Particulieren	1993	Combinatiepakket	40%
Avéro	1993	Comfort Plan	
De Noordhollandsche		Combipolis Particuliere	35%
Delta Lloyd	1989	Het Verzekeringspakket	
ELVIA Verzekeringen	1986	Alles-onder-één-dak polis	50%
	1974	Vier Sterren Polis	50%
Erasmus	1996	Multi Polis	40-50%
FORUM	1996	Voordeelpakketpolis	35%
General Accident	1993	Pakketpolis	50%
GENERALI	1976	Eén-Gezins-Polis	50%
Goudse	1996	Privé Woon Polis	
Guardian	1990	Comforta Polis	
Hooge Huys	1997	Totaalplan	30-40%
Interpolis	1997	Alles in één polis	almost 100%
Klaverblad	1997	Verzekeringspakket voor Particulieren	
Levob	1983	Privé CombiPlan	45%
	1983	Auto CombiPlan	77%
Nationale-Nederlanden	1992	ZekerheidsCombinatie Particulieren	
Nieuwe Hollandse Lloyd	1995	Odeon	60%
NOG	1997	Verzamelpolis	
Reaal	1993	Totaal Plan	25%
Royal Nederland	1985	Wel-Thuis Pakket	
	1985	Weg-Wijs Pakket	
Stad Rotterdam	1995	Autopakket	40%
	1995	Huispakket	40%
Tiel Utrecht	1995	Privé Pakket	
UAP-Nieuw Rotterdam	1994	Gezinspakket	50%
	1993	Uit & Thuis Combinatieplan	50%
Unigarant	1994	Pakket Verzekering	
Zevenwouden	1992	Particuliere Pakketverzekering	75%

Zürich	1996	Rubriekenpolis rondom het huis	50%
Zwolsche Algemeene	1985	Modulair Privé Pakket	85%
	1970	Auto Pakket Verzekering	85%

Source: Het Verzekeringsblad (VB), 24 April 1997.

More *integrated services* can be found in the directions of personal financial planning, personal or family risk management and flexible employee benefits and cafetaria plans. This supposes a pro-active research of the service provider into the formal and informal needs, attitudes and customer behaviour. Integrated customised service bundles do not only necessitate such an up-front analysis but also a dynamic and flexible perspective that goes together with a long-term client relationship. The degree of personalisation and flexibility can differ quite a lot as well as the type of products and services involved. There is a long list of possible combinations:

- banking accounts and payment services;
- savings and investment instruments;
- different types of loans and credit formulas;
- all types of private life and non-life insurance;
- risk management instruments other than insurance;
- social insurance covers and partly privatised categories;
- legal aspects (family, succession, contracts);
- etc.

Such an approach requires a great deal of individual information per client, a long-term planning in relation to the life cycle of the client with different event scenarios and a dynamic approach for flexible updating and adaptation.

Het Proveniersplan: certainty for the elderly

Again we refer to an example from the Dutch financial market. More particularly we focus on *Het Proveniersplan* of the Achmea Group which stands for certainty for the elderly. This product was developed in 1995 for all parts of the Group -the Achmea Group is a combination of different insurance groups and banks.

Servicing the segment of the older people will become a booming business. The opinion of Achmea is that products and services for this segment not always meet the needs of these people. The purpose of this *Proveniersplan* is to provide the elderly with products and services tailored to their needs so that they can maintain their current way of living as long as possible. This product is also the appropriate answer for the more general economic problems in the social security. More and more an appeal is made on the responsibility of the individual and contributions are asked for provisions made by the government.

The Proveniersplan consists of different elements and is about *integral home care*: the provision of an alarm system, help at home, security systems and meal preparation are all forms of servicing in kind. All the products of the Proveniersplan can be considered as variations or complements of existing products such as pensions, annuities, mortgages, general insurance, etc. The products will be adapted to the wishes and the needs of the elderly and are composed modularly so that everyone can make his own product.

For the moment the success of the Proveniersplan is still limited. The spokesmen of the different parts of the Achmea Group acknowledge that the new product is nothing more than a package of existing financial products and that for the products in kinds the Group is still in a testing phase. Furthermore the reactions on the new product are relatively modest so there is also a need for more promotion and publicity.

The *capital care* concept of OHRA

OHRA is a Dutch direct writer with a clear customer-oriented strategy. If we go through the annual reports we encouter terms like all finance, all care, and capital care. On average OHRA has 2 to 3 insurance policies per customer but it has figured out that the more prosperous clients have less contracts. OHRA has discovered that there should be a more appropriate communication of more adapted products on the right moment. As the client is supposed to take the decisions on his financial products himself OHRA tries to provide him with all the relevant information.

The capital care service was created to provide full financial advise. In this way OHRA strives for a lasting relationship with its clients by providing the client with professional help for banking, insurance, investment services. Thus OHRA wants to earn money *for* the client ! The financial group does this on a continual basis taking into account the particular situation and some characteristics of the client (e.g. the risk profile).

The starting point is the capital position of the client. This capital position is formed by an inflow, a capital position, and an outflow. If there is something left after the flows are taken into account then there is room for financial planning. A financial target is proposed based on the current capital position and on the risk profile of the client. All these factors determine which actions are taken and which products will be used. It is clear this is a very customer-oriented approach, as this is the basis for further actions. This is certainly not comparable with the selling process of a lot of other financial institutions.

3

LOOKING FOR THE RATIO BEHIND THE DIVERSIFICATION STRATEGIES

SOME GENERAL CONSIDERATIONS ON DIVERSIFICATION

The extent of diversification

In chapter 1, we argued that financial conglomerates, all finance, bancassurance and assurfinance can all be seen as ways in which companies in the financial sector implement a strategy of diversification. Mostly this diversification strategy is seen as an interesting way to realise a growth strategy in a mature market environment. However this diversification can also aim at the realisation of a competitive advantage through the exploitation of economies of scope and synergies.

The purpose of this chapter is to give a brief overview of the main topics in the diversification literature and to look whether the principles which hold in general are also applicable to the diversification issue within the financial services industry.

Diversification has been well-documented in strategic management, finance and to a lesser extent in industrial economics literature (Hoskisson and Hitt, 1990; Ramanujam and Varadarajan, 1989). Although there are a lot of variations in the definition, diversification in the broadest sense can be defined as the entering into new businesses by an existing business entity. As such, diversification can be seen as one alternative to achieve a growth strategy. This was noticed by Ansoff (1965) who considered diversification as a growth strategy by which a company tries to serve new markets with new products. However, Ansoff states that other alternatives are open to realign the firm's product-market environment. He has brought them together in his famous *growth vector matrix.*

"By its very definition diversification is the more drastic and risky of the two strategies, since it involves a simultaneous departure from familiar products and familiar markets" (Ansoff, 1965: 113).

Figure 3.1 Growth vector matrix (Ansoff, 1965)

Mission	Product: Present	Product: New
Present	Market penetration	Product development
New	Market development	Diversification

Generally a firm enters the diversification stage after it has observed that the current market does not guarantee sufficient growth perspectives. This process was described by Chandler (1962) when he studied the long-term development of large U.S. corporations. Chandler discovered several stages in this process of development:

- firms start as single product businesses supplying local markets;
- due to improved transport and communication possibilities the firm expands from a local market to a regional and even a national market;
- the next stage begins when firms grow by vertical integration, either backwards (acquiring supplying firms) or forwards (into marketing and distribution systems);
- excess capacity in marketing and distribution systems causes firms to diversify their product ranges (Grant, 1995).

Corporate diversification may take various forms. A common classification of diversification strategies starts from the characteristics of products and customers in relation to the present product-market position. Ansoff (1965) distinguishes between:

- *horizontal* diversification is a diversification strategy within the economic environment of the diversifying firm; this can either be done by using related or unrelated technologies;
- *vertical integration* occurs when a firm expands the scope of its activities to its suppliers or customers. As we just have mentioned Chandler discovered that vertical integration preceded the real diversification phase (see concentric and conglomerate diversification).

According to Ansoff, these diversification strategies offer only limited potential for the objective stated by the firm. Only if the present economic environment is healthy and growing, these strategies will contribute to a firm's objectives. The next two diversification strategies are more fundamental forms of diversification.

- *concentric* diversification is a diversification strategy into related products or services so that parallel technology and marketing can be used;
- *conglomerate* diversification is a diversification into completely different activities with a completely different technology and oriented towards different markets.

Rationales for diversification

Many motives have been forwarded to explain the reasons behind a diversification strategy. These motives cover a whole range, from efficiency enhancement to more troubling explanations where diversification is explained by managerial motives. In the latter case, there are no benefits for the shareholders. Therefore strategic management theory has tried to figure out under what conditions diversification is an efficient strategy, and in what circumstances it only benefits top management (by developing *empire building* growth strategies). We give an overview of the most frequently cited rationales for diversification.

Synergies and economies of scale and scope Perhaps the most cited reason why firms diversify is to achieve synergies by exploiting economies of scope. The synergy effect is often described as the 2 + 2 = 5 effect, which indicates that the corporate portfolio of businesses is worth more than the sum of the respective parts. This value creation is often ascribed to the existence and the exploitation of economies of scale and scope.

Economies of scale are a long-known concept, referring to the phenomenon that an increase in output can be realised with a disproportionate (lower) increase in input costs due to the decreasing effect of fixed costs and/or the beneficial effects of the learning curve.

Economies of scope can be defined as reductions in average and marginal costs that result from using facilities and processes in a single operating unit to produce a larger variety of goods or services (Maital, 1994).

In our opinion, these definitions are somewhat restricted in that they only focus on improvements at the cost level. However, an increase in revenues can also be considered as some kind of economy of scale or scope. In sum, what it is all about is that the production of different varieties leads to an increase in profits.

One of the most important books in this respect was written by Edith Penrose in 1959. At that time, her book *The Theory of the*

Growth of the Firm certainly did not have the impact it has now. The emergence of the resource-based view however has attracted people's attention to this remarkable study. The resource-based view argues that rent-seeking firms diversify in response to excess capacity in productive factors which are called resources (Montgomery, 1994). Penrose has stressed the importance of (excess capacity in) human capital. In her opinion this is the main driving force behind corporate growth and diversification.

The importance of synergies was also stressed in Porter's books *Competitive Strategy* (1980) and *Competitive Advantage* (1985). Business units that support and complement each other in their competitive efforts can have a value creating effect, due to the synergies between these entities. The hypothesis was that the more related the business units, the more synergies could be exploited. As relatedness enhances the level of synergies, it can be seen as a major potential source of competitive advantage. This idea originally stems from Rumelt who discovered a positive relation between diversification relatedness and performance (see later). Porter describes in detail how relatedness can contribute to competitive advantage. The two main approaches are (1) sharing of resources, and (2) the transfer of skills or expertise.

However Teece (1980) proposed that a cost function displaying economies of scope has no direct implications for the scope of the business enterprise (i.e. is no real motive to diversify). Only if economies of scope are based upon the common and recurrent use of a specialised and indivisible physical asset, then multiproduct enterprise (diversification) is an efficient way of organising economic activity.

Besanko e.a. (1996) describe a research project by Nathanson and Cassano (1982) who investigated how related multiproduct firms actually were. Their hypothesis was that:

> "If diversification is driven by a desire to exploit economies of scope, we should expect to see many more multiproduct firms with low levels of diversity than with high levels of diversity" (Besanko e.a., 1996: 224).

The results of their classification of 180 large multiproduct firms showed that many diversified firms were not positioned to exploit scope economies. Thus there must be other motives for firms to diversify.

Economising on transaction costs A second stream of literature focuses on the comparison of transaction costs versus bureaucratic costs. Transaction cost theory was developed by Williamson (1975) and since then has obtained an important position in economic literature. This theory regards the transaction as the basic unit of analysis and holds that the organisation of economic activity is largely to be understood in transaction cost economising terms (Riordan and Williamson, 1985). The central paradigm is that a firm can either use the market or internalise these operations within the firm. Suppose a manufacturing firm has the choice to buy its inputs from the market or to produce its inputs itself. Whether the firm buys or produces the inputs itself is dependent upon the amount of transaction costs, which are costs incurred when using the market, relative to bureaucratic costs, which are costs of co-ordinating exchange via the management hierarchy.

Under certain conditions using the market may be more expensive than relying on management hierarchy. According to Hill (1994) there are two main branches to the transaction cost literature.

- The first branch looks at *measurement problems* as a determining factor for the choice *make or buy*.

  "More generally, the measurement branch emphasizes that uncertainties as to the true value of the inputs that factors of production bring to an exchange, or uncertainties as to the value of the outcome of an exchange, inevitably give rise to contracting and monitoring problems" (Hill, 1994: 299).

  For example, if a bank decides to sell insurance products, then the bank may buy insurance products from the market or it may decide to product insurance products itself. However as the bank relies on the market and fears that some activities will not be performed well, a bank may decide to set up its own insurance company. In this way, the bank can have control on all aspects of the production and service process (from selling to claim handling) of insurance contracts.

- Managers may also opt for a hierarchical governance structure (i.e. within the firm) when there are a lot of *asset specific investments*. If transactions involve many non-specific redeployable investments, then there are not much incentives to internalise these transactions. On the other hand, as a condition of asset specificity becomes more important then firms will try to set up progressively stronger bilateral trading relations. Due to changing circumstances these relations may not always hold, and consequently firms are inclined to use more complex forms of governance. As Riordan and Williamson (1985) conclude: some transactions may be removed from the market and organised internally instead.

  The following case leads us to the same conclusions. A firm makes an asset specific investment, e.g. a company sets up a new plant to produce a product which will be used as input for one firm only (so there will be no other buyers for this product). As this buyer suddenly threatens to substitute this original input for another product, our firm faces opportunistic behaviour of this buyer. This can be considered as an ex post transaction cost and may also lead to the withdrawal of this transaction from the market.

Managerial motives for diversification: agency theory
The rationales we just described can be considered as efficiency enhancing motives, whether it is by exploiting economies of scope or by economising on transaction costs. Agency theory may

provide a third reason why firms diversify. Agency theory is concerned with the relation between a principal and an agent. Shareholders of the firm are considered to be the principals, while managers are seen as the agents. Managerial motives for diversification aim at maintaining or enhancing the positions of managers, rather than striving for value creation for shareholders. This situation may occur when shareholders are too dispersed to enforce value maximisation.

The same behaviour can be observed when the information asymmetry is high. When information asymmetry is high, managers are able to present information in a way it best suits the management's objectives. Diversification may then be pursued because their salaries and compensations may be linked to asset size or the corporate risk profile. Montgomery (1994) cites Mueller (1969) who stated that conglomerate mergers appear to be a convenient vehicle for doing so. In this way *empire building* managers rather choose to diversify into some unrelated industries than to give the free cash flow to the shareholders (free cash flow can be considered as cash necessary to fund all projects with a positive net present value - see also Jensen, 1986).

Besides this empire building theory, other theories are forwarded which can also be considered as part of the agency theory. Managers may choose to diversify from a defensive point of view and diversify in a way that the firm's demands for his particular skills increase (Montgomery, 1994). But often a firm diversifies beyond the optimum, and as a result shareholder value will be lost. This behaviour is called *managerial entrenchment*. This situation also occurs when a firm is the target of an acquisition. If the managers of the target firm fear for their position in the newly created firm, they may seek an alternative merger partner, hoping that this will secure their current positions in a better way.

This discussion is part of a broader discussion on the efficiency of the *market for corporate control*. It would take us too far to discuss this item in detail, but the general idea is that if the market for corporate control is efficient then there is a market-based limit on opportunistic managerial behaviour. It is still unclear whether this role of disciplining managers applies too the majority of the firms (Besanko e.a., 1996: 239).

Amihud and Lev (1981) have come up with another motive for diversification. They argue that managers may diversify in order to reduce total firm risk. Their job security is improved by the fact that they are no longer so vulnerable to the problems encountered in their specific industry.

Other motives for diversification There are other motives for diversification as well. We will describe them very briefly:

- *financial synergies*: diversification has sometimes been motivated to capture the potential financial synergies. Financial synergies arise from the pooling of imperfectly correlated income streams which results in a more stable and adequate cash flow to finance its activities. This is especially the case for unrelated diversification. This hypothesis holds as long as the capital markets exhibits failures so that investors are at a disadvantage compared to the diversified firm. Sometimes the possible effects of a merger on the tax liability of an acquiring firm is considered as part of the financial synergy motive. If a firm acquires a company with (huge) transferable losses, the firm may benefit from a tax advantage in that the overall profits have declined. Of course this motive is rather short-term oriented, and therefore it is of secondary importance and not a determining factor of mergers and acquisitions.

- *Market-power view*: sometimes (horizontal) mergers and acquisitions are aimed at reducing competition within an industry so that prices can be raised or costs for monitoring rival firms can be lowered. Regulatory authorities are very concerned about this motive and therefore carefully look for potentially anti-competitive effects. Indeed diversification may deliver a multiproduct company some advantages compared to specialists. For example, a conglomerate may cross-subsidise some activities at the expense of other activities. Mutual forbearance indicates that multiproduct firms compete with other multiproduct firms on several markets. When they see this is the case, they may be inclined to compete less severely with the other multiproduct firms and some form of oligopoly may arise.

Empirical evidence on the relation between diversification and firm performance

As the concept of diversification was described in the early 1960s by Ansoff, Grotto and Chandler, the boom of diversification literature started with the famous work of Richard P. Rumelt *Strategy, structure and economic performance* (1974). Rumelt updated and refined some of the concepts developed by Wrigley (1970) and was the first to find a relation between forms of diversification and firm performance. More exactly, Rumelt identified four main categories of firms based on their diversification strategy. Therefore he defined a firm's *specialisation ratio* (R_s) as the fraction of revenues accounted for by its largest single business unit. The *related ratio* (R_r) is the fraction of a firm's revenues attributable to its largest group of somehow related businesses.[2] To interpret whether a firm's businesses were related, particular attention was paid to the absence or existence of shared facilities, common selling groups, and other tangible evidence of attempts to exploit common factors of production (Rumelt, 1982: 360).

---

[2]Rumelt defined some other ratios such as *related-core ratio* and the *vertical ratio* but these ratios are only useful when we look at the more detailed classification Rumelt discovered.

Rumelt made a distinction between four main types of businesses (this is a commonly used simplification of Rumelt's more extended version in which he defined seven categories)[3]:

- Single business $R_s \geq 0.95$
- Dominant business $0.95 < R_s < 0.70$
- Related business $R_s < 0.70$ and $R_r > .70$
- Unrelated business $R_r < .70$

Rumelt was able to identify important performance differences for these categories. More particularly, he found that diversifiers which concentrated on some central skills or competencies (i.e. the dominant-constrained and the related-constrained category in his extended typology) outperformed all other categories. The other dominant businesses and the unrelated businesses had relatively lower return on equity (ROE). These results were very remarkable because they went against some generally accepted principles, and Rumelt was very conscious in interpreting the results:

> "The data, however, only indicated an *association* between these strategies and high levels of financial performance. While it may be that these strategies tend to produce good rates of growth and profit, it is more likely that firms that are already rapidly growing and profitable think it wise to restrict their scope of activity to business that directly relate to their currently successful areas of competence" (p. 156).

This study was the start of a huge stream of literature which refined and redefined the tools which were used by Rumelt. One of the main critics on Rumelt's classification was that the typology was difficult to apply in practice and subject to questions about reliability and researcher-specific classifications (Hall and St. John, 1994). Another frequently used measure of diversification was based on continuous measures such as SIC codes, but the drawback is that this measure fails to tap the dimension of relatedness.

The point is that the more studies were published on this issue, the more inconclusive the results. There were confirmations of Rumelt's 1974 study, but others have found opposite results (Seth, 1990; Lubatkin, 1987; Chatterjee, 1986). Studies in other areas (such as finance) have come up with the same conclusions. Besanko e.a. (1996) summarise:

> "three different lines of research on the performance of diversified firms have led to similar overall conclusions. Diversification can create value, although its benefits *per se* relative to nondiversification are unclear, due to industry effects and other factors."

---

[3] Dominant business consists of the following categories: dominant vertical, dominant constrained and dominant linked-unrelated; related business consists of related constrained and related linked business.

Among these factors are differing performance measures, diversification measures, samples and time periods (Lloyd and Jahera Jr., 1994). In paragraph 3.3 we will come back to some of the topics we have mentioned here.

These *inconclusive* empirical results enforced the contratrend of de-diversification, induced by shareholders who forced managers to stick to their core business and to dismantle the empires built with the free cash flow (see later).

The third merger wave

As we just have described most research shows a declining number of single business companies in the fifties, sixties and part of the seventies (see table 3.1). At that time, related and unrelated diversification grew significantly.

Table 3.1. Estimated percentage of Fortune 500 in each strategic category

	1949	1954	1959	1964	1969	1974
Single business	42.0	34.1	22.8	21.5	14.8	14.4
Dominant business	28.2	29.6	31.3	32.4	25.1	22.6
Related business	25.7	31.6	38.6	37.4	41.4	42.3
Unrelated business	4.1	4.7	7.3	8.7	18.7	20.7
Total	100.0	100.0	100.0	100.0	100.0	100.0

Dominant business consists of the following categories: dominant vertical, dominant constrained and dominant linked-unrelated; related business consists of related constrained and related linked business.

Source: Rumelt (1982)

The pattern shown in table 3.1 is generally considered to be the *third merger wave* and started in the late 1950s and the early 60s. This merger wave occurred primarily because regulatory authorities prevented horizontal mergers and vertical integration in order to encounter the aggressive antitrust enforcement. Furthermore, as the capital market was growing strongly buying shares (and thus taking over companies' stock) was a good investment. The large cash flows and high valuations of company stocks further stimulated the emergence of the large *conglomerates*. As firms were able to issue equity at attractive terms, they were reluctant to return these cash flows to the stockholders and invested it in growth and acquisitions (Shleifer and Vishny, 1991).

THE GENERAL TENDENCY TOWARDS DE-DIVERSIFICATION AND BACK TO THE CORE BUSINESS

Explanations for this strategic shift

The results of these diversification strategies were not very good. As we will see later, there was a large diversification discount during the 1960s, but this discount declined to zero during the 1970s (Servaes, 1996). Despite the long list of potential benefits, most research reveals a disappointing performance of diversified companies (and especially unrelated diversifiers):

- Porter (1987) revealed that more than 50% of all 33 large diversified firms he analysed over the period 1975-87 divested again due to disappointing results;
- Pekar (1988) found significantly high degrees of bankruptcy for diversified firms (formed through mergers and acquisitions);
- Kogut (1988) also observed significantly worser results for firms diversifying through joint ventures.

It seems that managers encounter many difficulties in the process of realising the diversification advantages. According to Collis and Montgomery (1995) the modern *resource-based view* on strategy helps to explain the poor track record of corporate diversification: (1) managers overestimate the transferability of specific assets and capacities while they mostly underestimate the impact of a specific market environment, and (2) managers overestimate their competitive power when entering competitive markets and in coping with entry barriers.

These new insights ran parallel with the *fourth merger wave* which started in the 1980s. This phase is characterised as a restructuring period because most researchers agree that the refocusing in the 1980s must be seen as attempts to reduce excessive diversification. The general adagio was back to the core business: researchers now agree that there is an optimal level of diversification and the merger wave in the 1960s was characterised by a diversification beyond the limits.

Another characteristic of this fourth merger wave was the emergence of the *hostile take-over*. Due to the financial innovations in the capital market (junk bonds), firms were able to take over large firms with a minimum of equity capital. These take-overs were labelled hostile because managers generally did not agree with the take-over. Nevertheless shareholders frequently agreed to sell their shares to the raiders (Besanko e.a., 1996).

Markides (1995) has given an overview of the reasons why conglomerates have underperformed their industry peers.

Researchers have found evidence that there were agency problems and that instead of returning cash flows to shareholders further acquisitions were made:

> "For example, studies by Baumol e.a. (1970); Brealey, Hodges, and Capron (1976); Gabrowski and Mueller (1975); Hiller (1978); and McFetridge (1978) -all of which looked at the *marginal* returns on capital- have found that the rate of return on investments financed by retained earnings is considerably lower than the rate of return on investments financed by new equity or new debt -and often lower than what shareholders could have earned by simply investing in the capital market" (Markides, 1995: 102).

Furthermore, diversification costs have increased over time compared to the benefits of diversification (e.g. by disclosure requirements). A third reason is that managers may have overestimated their capacities. This argument was also formulated by Collis and Montgomery (cf. infra). Last but not least, also the capital market is somewhat responsible for the over-investments in diversification. In the early 1960s, the capital market provided investors and managers with the wrong signals and incentives by reacting positively to announcements of unrelated diversification.

This relatively new view on unrelated diversification is also reflected in the opinion on *holding companies*. The mere existence of holding companies is under attack because its value added is questioned from different perspectives. In earlier days investors were looking for holdings as a mechanism to spread their investment risk. The last few years, there is a fundamental distrust from the investment society; this leads to a relative underperformance of holding companies, proven by the discount the stock market attaches to the equity value of holding companies. One can really wonder if the capital market does not make another mistake: is this underestimation economically rational or is it to be explained by distrust or misunderstanding?

However, the holding structure is found so complex that it is responsible for an opaque and unclear structure that also leads to extra (monitoring and control) costs. Furthermore, the reconciliation of the many divergent interests not only costs a great deal of management skills and time but often leads to a relative decrease in the competitive position over the long(er) run, due to sub-optimal allocation of scarce resources and cross-subsidisation.

Some examples of demergers and refocusing

The many potential fields for internal conflicts between the strategic business units involved can necessitate the splitting-up of the large conglomerates. Illustrative in this respect is the reasoning behind the recent intentions or even realisations in the breaking up of conglomerates such as AT&T, Allied Signal, GE, Hanson, ITT, Unisys, United Technologies, Westinghouse, ...

Case 1: The case AT&T

The most famous example has been the split-up of American Telephone and Telegraph (AT&T). The message that AT&T, the so-called largest and most powerful concern on earth, planned to break up the conglomerate into different parts (long distance, telephone equipment and computing, while selling of the capital arm) was received with great enthusiasm by the financial market partners. It was noticed that the parts would be worth more than 10 to 20% above the actual value of the conglomerate or would unlock shareholder value. The undervaluation of the conglomerate stock price could merely be explained by the fact that the investors' evaluation of a heterogeneous conglomerate is too much influenced by its less performing business units (AT&T GIS -ex-NCR- and AT&T telephone company), thereby ignoring to a large extent the benefits of business units with a high potential (the hardware department).

The group evolved in 1984 as a result of the split-up of the sole telephone company in a long-distance division (called *Ma Bell* or AT&T) and some regional companies (*Baby Bells*). AT&T is an exception among the world's telephone companies because it is also a leading supplier of telephone equipment. Some years ago it acquired the computer company NCR (now called *AT&T GIS*), which after all appeared to be a folly of CEO Robert E. Allen.

By splitting the concern in its three basic parts, it follows other well-known examples such as Imperial Chemical Industries, ITT, Hanson, GE and many others, and hopes to benefit from an increasing flexibility and a narrowed focus. As M.S. Forbes states it: "The information age puts a premium on speed for creating opportunities and for implementing up-to-date technology. Only such fast companies can achieve the productivity and create the products and services essential for staying successful" (Forbes, 16 Oct., 1995).

By breaking up the conglomerate, Allen hopes to solve the problems which originated in the specific situation of AT&T. As it was both a communications company and a supplier of communications equipment, there were sometimes conflicts of interest between the equipment and the services units.
The break-up allows the communications company to tackle the market very aggressively because the units won't need to be so considerate of one another any more. "We found ourselves always concerned about co-ordination and even reluctant to act sometimes for fear as we would hurt the other parts of AT&T", says Allen (Fortune, 16 Oct., 1995). The management of the integration was a very time-consuming activity but was not tenable any more.
The communications equipment division will also benefit from the split-up. As the link with the communications company will not be strong, the competitors might be expected not being too reluctant to buy the equipment of the AT&T subsidiary.

The break-up allows the managers of AT&T to narrow their focus. The minefield of strategic alliances AT&T had to pick through, was too broad and the managers had to deal with too many strategic issues at once.
The computer organisation GIS (Global Information Solutions) now has the opportunity to go its own way as a small independent computer company. It will not be very easy because GIS was the money-losing business unit of the

group. The restructuring of the group will cost 40.000 employees their job, twice as what was expected. However, the announcement of the break-up earned praise on Wall Street: hence the 11 per cent jump in AT&T's stock price.

On the Internet web site (D. Blankenhorn, 1996 - CMP Publications Inc.) this de-merger is welcomed because AT&T had become a FUBAR, which means *Fouled Up Beyond All Recognition*. Their point of view is defended as follows:

- "AT&T tried (for almost 12 years) to use its financial strength to dominate businesses it did not understand, most notably computing;
- AT&T suffered grievously from the *not invented here* syndrome: if something was not created by AT&T it could not be any good;
- there was a lack of market discipline: the demerger will force AT&T's managers to either make money or get out because every business unit will have to be competitive on its own."

Case 2: the case British Gas

One of the most recent cases is the demerger of *British Gas*, the largest corporate restructuring ever undertaken in the UK. The reasoning behind this demerger is interesting to observe (FT, 2 Feb., 1996).

- The first reason was to end the conflicts between the two business units. Three years ago, the Monopolies and Merger Commission already argued that the company suffered from an inherent conflict of interest, harmful to competition. The separation threw into relief that the conflicts between the two businesses were not just on the competitive front.
- A completely different management style was necessary: the trading business is an entrepreneurial activity, requiring sharpness of eye and fleetness of foot; the pipeline business is highly regulated, long-term and earns steady profits. The new structure should permit two very different companies to evolve: British Gas will become more entrepreneurial and the pipeline business more able to concentrate on its long-term goals.
- The management will have a stronger incentive to negotiate new terms since the losses will no longer be cushioned by profits from other activities.
- More costs have to be shaken out.
- From the consumers' point of view it could be good news: the demerger will increase confidence in the independence and encourage more competitors to enter the market. This will put extra downward pressure on prices; however this could mean bad news for shareholders (reduced profit margins and dividends) and competitors (more heavy competition, reduced prices and profits).

SYNERGY: A CONCEPT WITH MULTIPLE DIMENSIONS

Before these general business trends and their relevant literature can be confronted with the diversification in the financial sector, the synergy dimensions have to be studied in more detail.

As we have seen in the section on the relation between diversification and performance, the concept of synergies (and relatedness) has played an important role in this debate. From the moment researchers have focused on diversification, they have tried to operationalise the concept of *synergies*. Due to the strategic management theory, for a long time synergies were operationalised by looking at the degree of relatedness. Now researchers are convinced that this operationalisation is too uni-dimensional and that other aspects of the diversification process should be taken into account.

The concept of synergy and strategic fit (relatedness)

Considerable attention has been paid to the concept of *strategic fit* and synergistic benefits as determinants of acquisition performance (Datta, 1991). Originally, finance researchers have investigated whether results of acquisitions and mergers created value. The results were positive and value creation was ascribed to the presence of synergies between the acquirer and the acquired firm (Jensen and Ruback, 1983; Bradley, Desai and Kim, 1988). However, studies in finance have aggregated all types of acquisitions and thus have not investigated how managers increase stockholder value (Lubatkin, 1987; Datta and Grant, 1990). This question was tackled in strategic management research when the *relatedness* concept was introduced. Davis e.a. (1992) have defined relatedness as:

> "the degree to which business units support or complement each other's business activities, particularly marketing and production."

According to the authors, relatedness must be seen as the mechanism by which synergy is captured and thus is a major potential source of competitive advantage for the business units. This is also consistent with the resource-based view which postulates that managers deploy firm resources to markets they believe they lead to the most profit (Chatterjee and Wernerfelt, 1991). The more related the markets, the larger the profit potential due to the exploitation of economies of scope and the possibilities of transferring core skills between businesses (i.e. the leverage effect, which is very important in the resource-based view). Therefore, the hypothesis was formulated that related diversifiers should outperform unrelated diversifiers. In this way, the economies of related diversification are supposed to be larger than the economies associated with unrelated diversification. Empirical support for this hypothesis was provided by Shelton (1988), Davis e.a. (1992) and Brush (1996) who found positive performance consequences of exploiting synergies.

Not everyone agrees with this statement. For example, Seth (1990) argues that unrelated diversification exhibits other sources of value creation, e.g. coinsurance and financial diversification, which are less available in the case of related diversification. Thus, different types of acquisitions are associated with different sources of value creation, but no a-priori statements on the amount of value can be made. Seth's findings seemed to confirm this hypothesis: he found similar magnitudes of value creation for related and unrelated acquisitions. Similar results were found by Lubatkin (1987): he discovered that investors held the same expectations about related and unrelated diversification. He concluded that relatedness might represent an overrated concept. The findings of Chatterjee (1986) go somewhat further: Chatterjee analysed the impact of types of synergy on value creation. His conclusion was that the target companies in unrelated mergers fared much better than the targets in related mergers. However, part of these somewhat unexpected results were caused by the removal of the horizontal mergers out of the sample of the related firms. He predicted that inclusion of this category would have affected the results:

> "the related mergers as a portfolio would have performed at least as well as the unrelated merger portfolio and there would be no anomaly with the 'related is better' hypothesis" (1986: 130).

In general, it is still not clear whether strategic fit influences acquisition performance. As far as the financial services industry is concerned, it is argued that financial conglomerates have emerged to benefit from synergy opportunities because these are thought to be a major potential source of competitive advantage.

Relative size of the compounding parts

One of the most important moderating variables used when assessing the synergy issue is the relative size of the target company. It was hypothesised that acquisition performance was negatively influenced by the relative size of the target company. The main idea behind this logic was that smaller firms could use a more incremental approach -*learn as you go and expand the experience*. Now researchers found evidence for the hypothesis that the wealth gain of take-overs is proportional to the relative size of the bidder to the target (Chatterjee, 1986; Seth, 1990; Shelton, 1988).

For example, Hunt (1990) re-examined Kitching's results on a UK sample of 40 cases. His results confirmed Kitching's results on the importance of size mismatch for acquisition performance. Kitching discovered that a failure rate of 86 occurred if the target was too small (target's turnover less than 2 per cent of the buyer's turnover); Hunt's failure rate was less -67 per cent- but also significant. These findings are in line with the *critical mass theory* developed by Ansoff.

Kusewitt Jr. (1985) analysed the strategic acquisition factors which were related to performance. He found a more complex relationship between relative size and performance. When he first correlated performance measures with relative size, he discovered a negative relationship indicating that acquiring relatively large targets increased the risk to performance. However, when introducing a peaked quadratic function of the form $Y = a + bx - cx^2$ he found some evidence for an optimum relationship. Very small acquisitions seemed to be more troublesome because of the disproportionate management attention required relative to the small market contribution. On the other hand, larger acquisitions suffered from the *biting off more than you can chew* phenomenon both from a managerial and a financial standpoint of view.

The mode of diversification

The mode of diversification is a less-studied issue. However, there are studies which showed that the mode of diversification might be a relevant concept which must be taken into account when considering diversification strategies of firms. Generally distinction is made between internal diversification (internal development) on the one hand and mergers and acquisitions on the other hand. Internal diversification uses internal resources as a basis for establishing new businesses; mergers and acquisitions involve take-overs of target firms which must be assessed in terms of their strengths and weaknesses and the value they contribute to the acquiring firm (Datta e.a., 1991). Internal diversification is considered to be more profitable for some reasons: (1) internal diversification allows an incremental approach which is less risky; (2) it allows more flexibility because the firm can decide on different stages in the process to abandon unprofitable ventures; (3) there are less cultural clashes, which are more common in mergers and acquisitions; (4) there is a leverage effect that can be exploited, and (5) transaction costs are avoided.

Pitts (1980) tried to set up a contingency framework for diversification strategies based on previous research. He found that internal diversifiers strived for synergies by means of integration, while acquisitive diversifiers maintained relative autonomy for the different parts. This also had consequences on the organisational level: control systems were more strict and based on different performance measures in case of internal diversification. Furthermore, internal diversification was characterised by a larger corporate staff and more interdivisional transactions. Pitts's work was rather explorative in nature and needed some quantitative confirmation. This was done by Lamont and Anderson (1985), Simmonds (1990) and Busija e.a. (1997) who have investigated the link with corporate performance.

Lamont and Anderson (1985) showed that internal diversifiers had a significantly better return on assets than both acquisitive diversifiers and mixed diversifiers (who used both methods). However, the return on equity and the growth in sales were smaller. This made Lamont and Anderson conclude that acquisitive and mixed diversifiers might be posting higher sales growth figures than internal diversifiers at a price, namely higher financial leverage.

Simmonds (1990) found no significant differences between internal and acquisitive diversifiers, but when he simultaneously introduced the breadth of diversification (related-unrelated) he discoverd some interesting results. The related-internal category outperformed the other categories. Unrelated-external and unrelated-internal diversifiers showed the worst performance figures. Simmonds's results suggested that the combined classifications of diversification breadth and mode were more robust than either dimension alone, because the symbiosis in the decision-making process was included. Busija e.a. (1997) have re-examined some of the hypotheses of Simmonds and Lamont and Anderson and found performance improvements when strategy (breadth) and mode matched. However, this link was not stable in time. They also found that exclusive reliance on either internal development or acquisition could restrict performance. While a lot of scholars have focused on the distinction between internal development versus mergers and acquisitions, a growing stream of literature deals with (acquisitive) joint ventures. Joint ventures differ from mergers and acquisitions because they involve the setting up of a new company through the joining of two partners. It is noteworthy that joint ventures are seen as a somewhat short-term solution for a strategic problem. Of the six types of alliances, only one kind of joint ventures (*alliances of complementary equals*) are stable and successful (Bleeke and Ernst, 1995).

The need for organisational fit

A lot of researchers have stressed the need for *organisational fit* when investigating mergers and acquisitions. As the growing number of studies found no conclusive results on the effects of strategic fit, some researchers argued that strategic fit, while important, was not a sufficient condition for superior acquisition performance (Jemison and Sitkin, 1986). Datta (1991) stated that potential synergistic benefits will result in superior acquisition performance only if the synergies are realised through effective post-acquisition integration. In our opinion, the focus on organisational fit is consistent with the need for a more process-oriented approach as suggested by Hoskisson and Hitt (1990), Ramanujam and Varadarajan (1989), Nayyar (1992) and Hill, Hitt and Hoskisson (1992). According to Haspeslagh and Jemison (1991), the integration process is the key element in the whole acquisition process (1991: 105).

Of course, if the main purpose of the take-over is to achieve gains from financial diversification or from portfolio diversification, integration may not be necessary. Datta and Grant (1990) even showed that in unrelated mergers autonomy is associated with superior performance. However, when the firm tries to achieve operational synergies, then integration is important. From a theoretical point of view, the advantages of integration are clear: 'doing more (or better) with less'. But the real picture could be somewhat different from the expected one. The outcome of pursuing synergy is not always positive and sometimes 2 plus 2 can add up to 3 rather than the hoped-for 5 (Gupta and Govindarajan,

1986). According to Haspeslagh and Jemison (1991), managers acknowledge the importance of integration, yet it was the part of the acquisition process with which they were least comfortable. Thus, integration also involves costs which might be larger than the benefits, if not managed well. These *bureaucratic* costs are a necessary means to reap the synergistic benefits because interdivisional co-ordination is required. Furthermore these cost are higher the more interrelations there are, and thus, the higher the synergy potential (Jones and Hill, 1988). Davis e.a. (1992) point out that hidden costs and constraints may even be a more important negative factor because of the time needed to get all minds in the same direction and the inflexibility that is created. Nayyar (1992) distinguishes between governance costs, incentive depreciation and bureaucratic distortions but does not define the course of these costs relative to the degree of diversification.

Integration can be achieved on different activities or synergy dimensions. We focus on management integration, integration of other human resources, production integration, marketing integration and capital integration. Three of these synergy motives were also identified by Mahajan and Wind (1988). On the base of PIMS data, they were able to identify four synergy dimensions -sales (marketing), operations, investment and management- and linked them to performance. Their main conclusion was that synergy did have a significant impact on profitability but the magnitude depended on both the specific generating components and the type of business. Only management synergies had positive performance consequences for all industries. Mahajan and Wind did not explain their results. We try to come up with some theoretical arguments why each of these dimensions may influence performance.

Management integration may have implications on the strategic consensus of the top management team (TMT). It is clear that the more integrated the top management teams, the more these teams are confronted with each other and the more they are forced to look for a solution. When the top management teams remain relatively autonomous, diverging visions may evolve and co-operation may become more difficult. TMT strategic consensus has been studied over the years, but until now the findings on the antecedents and the performance consequences remain equivocal. Dess and Origer (1987) have reviewed the consensus-performance literature and concluded that the relation was not clear: greater levels of consensus were not directly linked with higher performance. Priem (1990) proposed a curvilinear relationship where the optimum level of consensus, relative to performance, falls somewhere between perfect consensus and perfect diversity of opinion within TMT. However, no empirical evidence was provided. A recent study by West and Schwenk (1996) analysed the above mentioned hypotheses but found no significant results. Management integration also has consequences on top management team heterogeneity. Where TMT heterogeneity was predominantly linked to functional capacities, we argue -like Barney (1986)- that different background and knowledge on different industries may be beneficial because of managerial synergy. Krishnan e.a. (1997) have proved empirical evidence for this hypothesis but they defined heterogeneity in terms of functional backgrounds.

The reasons for integrating marketing activities are more straightforward. *Marketing synergies* were identified by Porter (1985) when he described 'market interrelationships'. These interrelations may lead to lower selling costs or enhance differentiation so that the firm possesses a unique selling proposition. These kind of synergies were also identified by Rumelt (1974) and examined empirically in the paper and pulp industry by Davis e.a. (1992). Davis e.a. showed that firms characterised by high marketing relatedness achieved higher growth rates. However, this was not necessarily reflected in higher profitability. Mahajan and Wind (1988) discovered that -on the average- sales synergies resulted in a higher return on investments.

Marketing integration was used as one of the main reasons to form financial conglomerates. However, pure insurance companies or banks may have distribution agreements with partners from the other industry so that one need not form a group in order to benefit from marketing synergies. But as Riordan and Williamson (1985) have pointed out, as asset specificity increases (i.e. as a distribution channel has more value), internalisation (i.e. the formation of a group structure) may become more advantageous.

Production synergies are a third source of synergies and are also described in Porter's 'Competitive Advantage'. Perhaps this kind of synergy is less important to the financial services industry. There is still a legal prohibition for insurance companies to produce banking products and vice versa. Though, some insurance companies join their investment department with that of the bank and important synergies may be realised. This can be illustrated by the growing importance of an integrated asset liability management for the whole financial conglomerate. In general, Davis e.a. (1992) showed that production synergies can lead to higher profitability. Mahajan and Wind's conclusions (1988) were less clear-cut and revealed a mixed pattern.

Recently Rowe and Wright (1997) have theoretically proposed a link between diversification strategy and human resource management control. They argue that related diversification should be accompanied by strategic human resource management controls (as opposed to financial control). Furthermore, they argue that the implementation of strategic controls leads to flexible HRM strategies. Flexibility allows the firm (c.q. the conglomerate) to adapt more effectively to the changing environment. More integrated HRM departments should be able to deliver more flexible human resources enabling the conglomerate to react more quickly. Furthermore, there is a cost advantage in combining the two HRM departments, but this is of course very small compared to the total cost in a financial institution.

A last advantage which has been described in detail in many diversification studies are *financial synergies*. This kind of synergies seems to be the largest in unrelated diversification and occurs when the group acts as an internal financial market (Hill and

Hoskisson, 1987). In this way, external capital market failures can be more easily overcome and can give the business unit a competitive advantage if capital is a scarce resource. It must be mentioned that financial synergies are also available in related diversification. We argue that financial synergies are more easily achieved if capital integration is stronger. For example, if the capital ties between two firms are small, the potential for achieving financial synergies is much smaller than if two firms are 100 per cent subsidiaries of the same holding company.

Resource relatedness

In the sections described above we have argued that the concept of strategic fit is not sufficient in explaining performance differences between different types of mergers and acquisitions and different types of diversification strategies. We have shown that there were moderating variables such as relative size and mode of diversification. Furthermore we have shown that organisational issues are at least as important when it comes to the success of a diversification strategy.

At a more general level, the concept of relatedness has been reconsidered by proponents of the *resource-based view*. This new strategic stream is thought to be as important in the 1990s as industry analysis was in the 1980s (Collis and Montgomery, 1995). According to these authors, the resource-based view is grounded in economics and tries to find an answer to the question how a firm's resources drive its performance in a dynamic competitive environment. So both an internal and external perspective are used. However, the stress is put on the internal analysis of the firm. More specifically, the resource-based view to strategic management focuses on costly-to-copy attributes of the firm as sources of economic rents and, therefore, as the fundamental drivers of performance and competitive advantage (Conner, 1991). Firm's resources should have some characteristics in order to contribute to competitive advantage: (1) resources should be valuable, (2) resources must be rare, (3) they must be imperfectly imitable, and (4) there cannot be strategically equivalent substitutes for this resource (Barney, 1991).

Much attention has been paid to the diversification issue and attempts are made to explain diversification from a resource-based approach. This resource-based theory also adopts the view that related diversification will result in higher rents than unrelated diversification because of the greater likelihood of synergy. However, only if the synergies achieved are *idiosyncratic* and *bilateral* then competitive advantage will be reached (Mahoney and Pandian, 1992). Put differently, only if both partners benefit and if the sources of synergy are specific to the partners involved (i.e. are not easy imitable) then diversification will result in higher performance. The researchers have tried to define relatedness more creatively, away from the traditional approach based on product-market similarity. According to Markides and Williamson (1994) the traditional way of measuring relatedness between two businesses is incomplete because it ignores the strategic importance and similarity of the underlying assets residing in these businesses. Furthermore the authors state that the traditional way of operationalising relatedness is too static and does not take into account that relatedness also increases the potential for the firm to expand its stock of strategic assets and create new ones more rapidly and less costly. Thus more and more researchers have tried to incorporate these elements in the concept of relatedness. However this new approach is still emerging and until now there is still a lack of uniformity in the concepts and constructs used.

ALL FINANCE AND FINANCIAL CONGLOMERATES AS A DIVERSIFICATION STRATEGY: A PARADOX?

The diversification strategy of financial firms confronted with the larger business world

Looking at the business practice, one can observe that many changes on the level of companies and markets are not unique or isolated, but seem to fit into more general business trends and developments, like mergers and take-overs, internationalisation, business process re-engineering, TQM, etc. If this is the case, much can be learned by looking at other sectors, where trends may have emerged earlier, or at a broader and eventually more visible level. This comparison will certainly help us to ask the right questions and determine the most relevant issues.

Looking at financial conglomerates in this way creates an interesting *paradox*: while financial firms are forming conglomerates, many large firms are refocusing on their core activities, (industrial) organisations are down-scaling, refocusing to improve flexibility and are even breaking up (by MBO or otherwise).

It is therefore of the utmost importance to investigate whether this general business trend is in reverse of the integration in the financial sector. Is it possible that both types of diversification are not comparable or is it because the financial sector is witnessing a time lag so that it is yet in a phase of integration which the more industrial colleagues witnessed some decades ago?

Diversification into related fields

In the former section we have pointed out that the main adagio in the business world is back-to-the-core business. If core business is defined in a broader sense than is traditionally the case, then related diversification might be the appropriate strategy. Our hypothesis is that financial conglomerates are not a *conglomerate* diversification as defined in the (industrial) management literature but they can be considered as a related diversification. As a consequence, it is not necessarily a contra-indication of the general business trend.

There is a difference in timing which can explain the (seemingly) opposite directions. The trend of unrelated diversification is partly explained by the strict antitrust regulation, which made it impossible to grow by merger or acquisition of related companies. Once those laws were changed or adapted in a more liberal way, companies started to refocus by unbundling (de-diversification of unrelated activities) and rebundling (into the direction of more related diversification). Even the financial sector is witnessing this trend. It was not until recently, that companies in the financial services sector were allowed to merge with or acquire companies in other sectors at all.

In the light of the older, more general trend and its experiences, it is likely that financial service providers will diversify into related areas. One way of doing so, is by forming a financial conglomerate. In fact, it seems that the financial sector is moving towards the same types of strategy, but that it is coming from a different background.

The concept of synergies translated to the financial sector

In the general part we have shown that motives for diversification can be either value-creating or managerial. In this paragraph we will only consider the value-creating rationale for setting up a financial conglomerate. More particularly we will analyse if economies of scale and scope are present and if there is room for exploiting synergies.

Researchers as well as practitioners raise doubt as to the existence of *economies of scale* within the financial sector. Nevertheless it is clear that the needs of the modern financial market environment increasingly lead to a higher potential for sources of economies of scale:

- because of the growing importance of fixed investments (information technology, advertising, product development, marketing and management information systems, etc.);
- because of the mere existence of the need for a learning organisation.

As already mentioned *economies of scope* exist if the joint production cost of producing two or more outputs is lower than if the products are produced separately. Such advantages can relate to the comparable scale advantages, to a better risk spreading, a more optimal use of core competences, etc.

Most of these advantages will lead to a lower cost level and to a more reliable profit level and a more stable cash flow. Another important source of potential economies of scope is offered by modern technology, which makes it possible to produce different types of products and to deliver tailor-made products and individual flexibility (scope) with the scale advantages of standard single product lines.

> "Economies of scope exist where it is cheaper to produce in combination than separately, where it becomes possible to combine efficiency and flexibility (previously mutually exclusive), where it becomes possible to expand the product life cycle (to innovate) and diversify into other products. As such the fundamental assumptions are totally different: variety instead of standardisation, multi-product forms instead of specialisation, flexibility towards market changes, decentralisation instead of centralisation" (Van den Berghe, 1989).

These advantages will probably lead to a better competitive advantage which allows the firm to attract a greater market demand (see in this respect the success of concepts such as all finance and personal financial planning, personal risk management and all care, but also mass customisation).

The subject of economies of scale and scope has had a great appeal to the scientific society. In order to test the societal value of the financial conglomeration trend an interesting research has been ordered by the OECD (Forestieri, 1993). This research has surveyed all relevant studies (for the banking, insurance and securities business) over the period 1962-1989. Although this period does not incorporate the increased emphasis on *industrialisation* and *diversification* of those firms, it is the most relevant review up to now. These are of the most striking results.

- There exist many differences in relation to the time periods studied, the countries covered and especially the methodology used.
- Sources of scope and scale economies can be found in the following directions: technology (not only the relative importance of technology but especially the volatility of the hard- and software involved creates potential for economies), specialised professional skills, information as a production factor. Besides these supply-driven advantages also a number of demand-driven stimuli can be observed: e.g. the clients' need for integrated services, one-stop-shopping, global customer care etc.
- Technology can however lead to an increased flexibility so that the minimal scale levels can be lowered.
- There is no unanimity as to the existence of economies of scale and scope in the different types of financial firms. Economies of scale in insurance were frequently found but the optimal level was situated at a rather low level and could not explain significant advantages for larger firms.

One may however not forget that many of these results go back to a period during which the diversification towards financial conglomerates was not important at all, while the factors stimulating those economies were far less important than they are today in a highly competitive, technology-driven buyers' market that has lost most of its regulatory barriers. Another reason why potential advantages of diversification could be underestimated is that most research solely focuses on cost advantages, whereas numerous advantages are situated on the demand side and must be measured in quality level, client retention, etc. The classical approach for measuring scope and scale economies is to look at potential effects on historical cost levels, whereas many of the diversification strategies are anticipating upon improvement of the future market chances and of the competitive position.

The impossibility to distinguish between the effects of scale and scope leads to the suggestion to incorporate all those potential benefits under the common denominator of *synergies*.

Some sceptics doubt whether synergies exist at all in the financial sector and claim that the potential synergies are too difficult to achieve. Although there is little doubt that the diversification of financial firms into each others territory is mostly of the related type, quite a number of market partners and researchers still question the usefulness and long-term character of these strategies.

This opposition against financial conglomerates and all finance can certainly not be explained (alone) by a reluctance towards offensive innovators or a fear of costly diversifications. But sometimes criticism is expressed as a way of defending a relative competitive disadvantage (e.g. because no partner could be found or a potential joint venture did not materialise).

However, even the most enthusiastic defenders of the all finance movement and the financial conglomeration trend cannot neglect that some obstacles can play an important role in realising the potential benefits of the many types of diversification strategies. But here again, we can refer to the general business theory which described similar findings (see paragraph 3.3).

Is big (and integrated) still beautiful?

The new strategic thinking focuses on flexibility, speed of adaptability and offensive and creative innovations. Such a strategy is supposed to have a greater chance of success with smaller firms who upgrade through networking than it is for larger and global players, who suffer from bureaucratic inflexibility and defensive attitudes.

In the general business environment the move towards refocusing and de-diversification is complemented by a revival of the importance of networking, joint ventures and virtual organisations. The business practice is following the strategic recipes of the many management gurus and consultants (or is it vice versa?):

- "what is increasingly clear, ... is the fact that only an extremely focused unit (somewhere between 30 and 200 employees ...) functioning in a network more or less loosely co-ordinated by a 'server' can be quick and flexible enough to survive and prosper in the market place" (Hinterhuber e.a., 1994);

- the advantages of a network of smaller or mid-size firms are straightforward: lean, fast, extremely flexible;
- a network has the advantage that it is constantly possible to correct, adjust or change the network;
- networking offers the same scale advantages as larger competitors;
- it is much easier to build up a network from small parts than to break up a larger block into small parts.

According to Hinterhuber e.a. (1994) this strategic networking is the logical step in the evolution process of the business structure: starting with a non-controlled chaos, evolving into old-fashioned conglomerates, refocusing and re-engineering afterwards, to arrive finally at strategic networking. This networking strategy is however not a straightforward success. In the past many joint ventures failed. Stafford (1994) explains this bad experience with the fact that the partners involved expected too much too soon. Only after the partnerships mature (i.e. when the operations of the alliance are well-established and well-understood), managers can gradually shift to measure output, such as profits and cash flows. Immediate use of output measures are appropriate only when partnerships are engaged in industries with short product life cycles, or are designed for the short term or to realise cost reduction goals. Another element of disturbance can be the fact that networking partners sometimes have a *schizophrenic* relationship: they may co-operate with each other in one value chain, while competing aggressively in other markets. Moss Kanter (1994) highlights especially the bottlenecks that may occur on the personal level:

> "too often, top executives devote more time to screening potential partners in financial terms than to managing the partnership in human terms... they tout the future benefits of the alliance to their shareholders but don't help their managers create those benefits... they worry more about controlling the relationship than about nurturing it... they fail to develop their company's *collaborative advantage* and thereby neglect a key resource."

The integrated strategy behind (large) financial conglomerates

The creation of financial conglomerates, again, shows an opposite evolution. It is argued that the benefits and opportunities, the so-called economies of scope of a financial conglomerate, are higher the more organisations are integrated. Often the hypothesis is formulated that the potential synergies can only be exploited (to the fullest) when an integration strategy is followed. Several sources have come up with this idea. For example, Coopers and Lybrand (1993) have surveyed the largest financial conglomerates in six European countries and found that institutions that recognise or expect the most benefits and opportunities from all finance integrate the furthest. Maybe the fact that greenfields (start-up from scratch) seem to be more successful than mergers and acquisitions could also prove the necessity for far-reaching integration and co-ordination. The same reasoning could be found in Lafferty Business Research (1994) who stated that the main success factors do not, in themselves, guarantee results. And a manager from a financial conglomerate expressed it this way:

> "We do not believe in non-committal, finite co-operation agreements between a bank and an insurance company, when the object is to maintain or acquire a substantial market position" (Holsboer, 1995).

Thus, central co-ordination and control will also be necessary in order to cope with potential conflicts of interest.

Many financial conglomerates focus on integrating distribution facilities. Some organisations aim at a much higher level of integration. There seems to be a link between the degree of integration and the experiences gained; the longer institutions are involved in this diversification process, the further they integrate. Coopers & Lybrand (1993) distinguished three levels of implementation within financial conglomerates:

- aiming at increasing the sales effectiveness and customer base utilisation;
- looking for an improvement of the efficiency of support and processes;
- optimising the development of skills and the allocation of assets and resources.

Their suggestion was that integration should follow the value chain backwards, thus beginning with sales and marketing and expanding to other business functions afterwards. These concepts will be the base for the *financial conglomerates control board* (F.C.C.B.) which will be discussed in chapter 4.

4

THE FINANCIAL CONGLOMERATES CONTROL BOARD

TOWARDS A CLASSIFICATION OF FINANCIAL CONGLOMERATES

One of the remarkable elements of the trend towards financial conglomerates and all finance is the huge *heterogeneity* in this development process. It would be a mistake to consider financial conglomerates as a set of homogeneous groups of financial firms with similar characteristics. Some researchers have acknowledged this heterogeneity and have developed a typology of financial conglomerates. In the next paragraph, we will discuss some of the typologies we found in the financial literature.

Typology based on the corporate structure of the financial conglomerate

A first typology of financial conglomerates was developed by Herring and Santomero (1990). Their main concern was that U.S. banks would increasingly face competition from foreign financial conglomerates -foreign because in the U.S. there are still regulatory barriers between the different sectors in the financial services industry while in most other countries these barriers were relieved. Herring and Santomero fear that these financial conglomerates will have a competitive advantage because they are able to select from a broader menu of corporate structures to provide a wider range of services. In this way, these financial conglomerates may -theoretically- benefit from the economies of scope and as a result they may have an advantage over *specialised* U.S. banks. Herring and Santomero argue that there are diseconomies of scope as well, e.g. (1) bureaucracy and (2) complexity of managing several kinds of businesses. Therefore a full integration of a bank and an insurance company may not be optimal. By choosing for an in-between solution financial conglomerates may try to achieve the benefits while avoiding the diseconomies.

In the light of this discussion, Herring and Santomero have come up with alternative models, based on how a financial conglomerate has organised its corporate structure.

Model 1: The German model - complete integration In this model, the managers are permitted to conduct all activities within a single corporate entity. Although complete integration may result in superior operating efficiency, the managers of financial conglomerates may advocate some separation in the form of Chinese walls. This means that although there is no legal obligation to separate particular functions, the managers may decide to do so because this enhances the perceived value of their services to potential customers. For example, some firms will not be inclined to ask for credit facilities if a bank has some equity capital in it because the bank may make abuse of the information she has from her position as investor. However if the bank places some Chinese walls (firewalls) between the investment and the credit department, potential customers may be reassured and may ask for additional services from the bank.

As this model is perhaps the most attractive to the financial conglomerate, for supervisors this might not be the case. The regulatory authorities are challenged to understand the prudential implications of a large number of different products and this may result in costly and imperfect regulations.

Model 2: The British model - bank parent, non-bank subsidiaries In this case, there is a legal distinction between the bank and the non-bank subsidiaries. In this way there is less room for operational synergies. On the other hand this model protects the bank from a disastrous outcome of the non-bank operations. However, income streams flow back to the bank and in this model the non-bank subsidiaries can only increase but not reduce the value of the bank. Herring and Santomero acknowledge that these advantages are somewhat theoretical because in case of a failure the bank will also suffer due to the reputation effect (or the risk of contagion).

Model 3: The U.S. model - holding company parent, all activities as subsidiaries This model differs from the second model since the bank is considered as a subsidiary of the holding. Herring and Santomero argue that in this model the legal distinction is more extensive so that the potential for exploiting synergies is more limited: synergies can be achieved in the marketing and distribution function.

This model is preferred by supervisory authorities because this structure is more transparent. Furthermore, because of the indirect relation, the safety and soundness of the bank can be more isolated from the non-bank activities and the bank may be less inclined to bail out a company in trouble (Koguchi, 1993).

Model 4: Holding company parent, complete operational distinction In this last model the holding company operates as an investment company and no operational synergies between the different parts are exploited. Herring and Santomero conclude that this model should not be considered as an integrated financial corporation. This structure only benefits from the financial synergies which originate from diversification of income streams. Of course this model yields only limited supervisory problems.

Comments on the typology of Herring and Santomero The typology of Herring and Santomero is interesting because it points out that differences in corporate structure may have implications on the profit potential but also on the risk level of the financial conglomerate. There are however some drawbacks with the typology of Herring and Santomero. First, this classification is rather theoretical and is of less practical relevance to managers of financial conglomerates. It is worthwile to mention that when insurance companies are involved in the analysis, the German model simply does not exist. Even in the European countries where almost all barriers between the different subsectors in the financial services industry are relieved, production of insurance by banks and of banking services by insurers is generally prohibited in all countries (Hoschka, 1994: 8). Of course the main purpose of the authors was to review the U.S. policy regarding financial conglomerates and in this respect regulatory issues might be of more importance than developing a classification which is useful for managers.

Another drawback of this study is that there are more alternatives to form a financial conglomerate. Some of these groups have opted for rather tight relationships and have merged with, acquired or set up an insurer or bank. Others have chosen looser forms of cooperation such as joint ventures and marketing agreements. These latter forms were excluded from the analysis.

Herring and Santomero also strongly focus on the bank situation and describe models where banks are dominant. However there exist a lot of financial conglomerates where the insurance company occupies the dominant position. Thus this typology is limited to part of the financial conglomerates' sample.

A last drawback is the authors' focus on legal distinction as the criterion for their classification. Although the authors have distinguished between legal distinction and operational distinction, legal distinction is the main determinant for potential synergies and risks. Empirical evidence shows that nowadays financial conglomerates have set up corporate structures which have nothing to do with legal structures. It is this new corporate structure which determines the potential for economies of scope and the risks involved. Therefore the single use of legal distinction may be misleading and may be based on false hypotheses.

The tripartite subdivision (Verzekeringskamer, the Netherlands)

The second classification we propose was developed by the Verzekeringskamer, which is the Dutch government institution responsible for the control of the Dutch insurance companies. This classification is used to control financial conglomerates effectively and in this respect, it has some practical relevance (for the supervisors).

A distinction is made between:

- primarily banking financial conglomerates;
- primarily insurance financial conglomerates;
- mixed financial conglomerates.

In this way, financial conglomerates are classified by the proportion between banking and insurance activities. The Verzekeringskamer uses the balance sheet total as a criterion to distinguish between these categories. Other criteria should be used because balance sheet total distorts the results in favour of the credit institutions because it contains all monetary transactions with clients as well as in relation to inter-bank funding. For insurance companies these transactions are located on the profit-and-loss account. Van den Berghe and Verweire have suggested to base the typology on the type of the parent company. Then, a primarily banking financial conglomerate is a conglomerate with a bank parent, a primarily insurance financial conglomerate has an insurance company or an insurance holding as a parent and a mixed financial conglomerate consists of a holding company with both a banking and an insurance part (Van den Berghe, 1995).

Another possible criterion could be the turnover. Therefore, one should know which part of the turnover is created by banking activities and which part is procured by insurance activities. On the basis of this division, financial conglomerates can be assigned to one of the three groups. But again, the results are distorted: the insurance companies are favoured because their turnover figure gives the complete monetary transactions, whereas the turnover of banks only integrates the interest income of monetary transactions. Furthermore, it is not always possible to detect which part of the turnover is attributable to banking or insurance activities if the investigation is based upon annual reports. Moreover the more integrated the products become, the more difficult and irrelevant it will become to make the distinction.

Typology based on the method of entry

Some early publications (Lafferty Business Research, 1991 and 1994; Hoschka, 1994) have set up a typology based on the entry strategies of the banks into the insurance industry. Of course this typology may also be applied to insurance companies who enter the banking industry. This typology is interesting because it also tries to provide an explanation for success of some financial conglomerates.

Generally these publications make a distinction between:

- de novo entry;
- merger/acquisition;
- joint venture;
- distribution alliance.

Figure 4.1 Alternative entry vehicles for banks entering insurance (Hoschka, 1994)

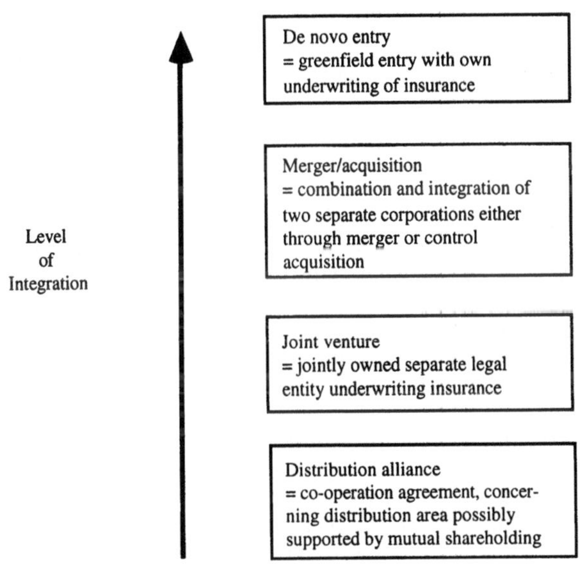

In chapter 1, we have described these different strategies into more detail. Here we only give a short overview of the main characteristics.

De novo entry As we can deduct from the figure, a de novo entry is characterized by the highest degree of integration. And as a matter of fact, there are some examples such as TSB in England, Crédit Agricole in France and Deutsche Bank in Germany, who proved that their *de novo entry* in insurance has been successful.

The relative success of this method of entry is often explained by the fact that with a de novo entry, the strong cultural differences can be overcome more easily. The banks (c.q. the insurance companies) have a higher degree of control over the whole start-up process and they don't have to take into account an insurance (c.q. bank) partner with diverging ideas and a totally different culture. This is an aspect that should not be underestimated.

Mergers and acquisitions According to publications of Lafferty Business Research (1991) and Hoschka (1994), mergers and acquisitions are the next best option for banks and insurance companies.

- Both a merger and an acquisition have the advantage that it is easier to follow one direction in the all finance approach. In case of an acquisition this is enforced by the parent company who has the power to outline the business policies; in case of a merger, the bank and the insurance company also have similar ideas about the most appropriate way to follow.
- The expertise and experience needed in the banking and the insurance domain is available in the group from the start.
- Furthermore, if the brand name of the subsidiary is well-established, the parent company benefits from keeping that brand name.
- This method is also suited when a quick entry is necessary.

Here again, some examples indicate that banks can be in a better position to exploit these advantages when acquiring a majority holding in an insurance subsidiary than is the case vice versa. Besides the problem of finding a good partner, there is another major drawback: mergers and acquisitions are also very expensive. This is certainly the case for an acquisition, because the price to be paid is mostly higher than the intrinsic value of the company.

Joint ventures A joint venture is defined here as a kind of a co-operation agreement between two or more independent companies, setting up a legally independent entity, owned and controlled by the parent companies (here a bank and an insurance company). A joint venture is considered to be a more formalised way of co-operation.

We have already argued that joint ventures are rather short-term solutions because nearly 80 per cent ends up in sale (Bleeke and Ernst, 1995). Nevertheless this is a frequently used entry vehicle, especially in cross-border alliances. The most cited reasons for setting up a joint venture are:

- the avoidance of costs of setting up a new business;
- the sharing of risks and costs with more partners: as a consequence, one does not need as much resources as with the other two possibilities;
- to appeal to a partner in order to gain knowledge and experience about the new business.

Distribution alliances - marketing agreements

> "The most common vehicle for bancassurance, marketing agreements allow firms to test the water before plunging into more expensive forms of co-operation" (The Economist, 1990).

Marketing agreements generally refer to distribution agreements: banks distribute insurance products via the bank branches while the insurance companies sell savings, investment and/or credit products through their sales channels.

It is a loose form compared to the other approaches and therefore it is at the bottom of Hoschka's figure. However, marketing agreements can be completed with substantial cross-shareholdings.

Several reasons can be at the root of this choice: this method of entry gives two (or more) companies the highest degree of freedom. We think this is rather a disadvantage because in my point of view, they don't get the most out of this relation. The combined offering of insurance and credit services requires a lot of involvement to resolve the bottlenecks that surely will appear. With this method, the two parties will give up efforts more quickly and will pay attention to other priorities.

Comments on the method of entry typology This classification is an attempt to explain differences in performance of financial conglomerates.

> "A comparative analysis of the different entry vehicles shows that of the four different entry routes, de novo entry seems to be most successful. This stems primarily from the organizational and strategic flexibility which de novo entry offers, allowing a tailored solution to the ideosyncratic internal structure and organizational environment of the bank" (Hoschka, 1994: 137).

However, we have already argued that the method of entry is only one element of the diversification process and that other aspects should be taken into account. For example, it does not consider other degrees of integration, such as integration on the product level, or integration on the organisational level.

Furthermore, this classification does not capture the dynamic aspect of the relationship between financial institutions. Coopers & Lybrand (1993) have shown that the implementation of bancassurance involves different stages. Only if this study is repeated each year, it is possible to observe shifts in the structure between these relations.

THE FINANCIAL CONGLOMERATES CONTROL BOARD

Definition of the financial conglomerates control board

Because of the limitations of the typologies we described, we thought it might be interesting to develop a typology which incorporates the comments we made.

The profound variance in the diversification strategies and structures, the difference in the many products and markets involved, the distinction in the level of integration are all factors that make it impossible to analyse these potential synergies without a standard reference framework. To our knowledge, such a reference base is not available yet.

We have tried to elaborate a useful *theoretical framework*, by combining the concept of the value chain with the service operations system (the servuction system). This standard reference base, called the *financial conglomerates control board* makes it possible to gain a better understanding of the types of financial conglomerates, their structure and strategy, their potential synergies as well as the types of potential risks.

Figure 4.2 The financial conglomerates control board

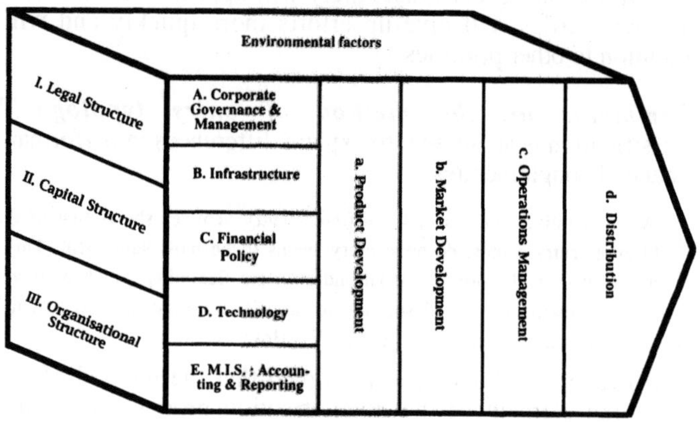

The concept of the *value chain* has been developed by Porter. The value chain highlights the different activities to create added value for the customer. Every step costs money and creates value. The higher the total added value, the better the perspective for a competitive advantage and a profitable margin. This tool is mainly used for an industrial firm. In order to be able to integrate this concept into the analysis of financial conglomerates a fundamental adaptation is necessary. Such a translation can best be based on the structure of the service operation system, which pictures the banking and insurance firms. The details of this framework must be so that they make it possible to test for:

- the type of conglomerate from a legal perspective;
- the method of entry used;
- the type of implementation, collaboration, co-operation and integration;
- the potential types of synergies;

- the potential risks involved (business risks as well as societal risks).

The reference scheme we developed is therefore far more detailed than most previous attempts (see e.g. van Rüth (1994) and Coopers & Lybrand (1993)).

The goal of the instrument

The ultimate goal we aim at with this instrument is twofold.

- It will be a reference frame for further research. At first this research will be of a qualitative nature; later it will be refined with more quantitative data.
- We hope this instrument will become a tool to be used by different parties involved with financial conglomerates. The aim is to develop it as a tool to control and evaluate:
    - the drivers for and the different types of synergies within financial conglomerates, and the degree to which they are realised;
    - the drivers for and the different types of risks within financial conglomerates, their size and possible consequences;
    - the classification of financial conglomerates and possible correlation with the performance level. In the future we hope to classify conglomerates according to differences in this reference framework. In addition the specific macro/market environment is also relevant and should be considered as well.

Detailed description of the financial conglomerates control board

This reference scheme is divided into three main parts. The arrow on the front side (containing the boxes *A. Corporate governance and management* to *d. Distribution*) is comparable to the value chain. It depicts all *activities* where synergies can be found and which determine the risk level of the conglomerate. At the *back side* of the instrument, we find some structural aspects (such as the legal structure, the capital structure and the organisational structure) which also influence the synergy potential and the risk level. At the *upper front*, we take along some environmental factors. These are control variables which could also be important in the comparative study of financial conglomerates.

Let us start by explaining the structural factors in somewhat more detail.

- The legal structure is the way in which two (or more) companies are formed. The early studies on financial conglomerates suspect a relation between the degree of success and what they call *the method of entry*. We can use the same subdivisions Hoschka and Lafferty Business Research made (see point 4.1.3). It is commonly accepted that start-ups and mergers and acquisitions lead to better results.
- The capital structure gives insight into the capital ties between the different companies of the group. One of the most important risks, namely the risk of contagion, is to a large extent of a financial nature. This means that the stronger the financial ties between different companies, the greater the potential for contagion.
- Sometimes financial conglomerates set up a new corporate structure which should facilitate co-operation between the different parts of the group. However not all financial conglomerates adapt their organisation structure but stick to the legal structure. It would be very interesting to investigate if such an adaptation leads to better results.

The *activities* are the main part of the financial conglomerates control board. As in the value chain of M. Porter, we have made a distinction between primary activities and support activities. The primary activities are indicated with a small letter, the support activities with a capital.

The *support* activities can be further subdivided into several categories:

- Corporate governance and management: we have identified some relevant issues in this respect. As management is a very important aspect in all these matters, we have tried to identify which management structure is best suited to influence the synergy process. Furthermore, we will investigate whether this has consequences on the strategy process and the human resource management. We will also analyse the impact of synergy development and cultural projects on the diversification process.
- It is a question whether a physical integration (infrastructure) is an essential element in achieving synergies.
- From our case studies we have seen that financial conglomerates try to achieve synergies in the investment department. The issue consists of different elements: we will investigate whether the investment policy is integrated, whether there is an integrated asset liability management, whether there is an integrated risk management and whether reinsurance activities are grouped together.

- We also want to analyse the technological issue. To what extent are technological systems of the different parts of the group used in other parts of the group and to what extent do they fit?
- When a financial conglomerate is created, it is interesting to see what implications this decision has on the management information system of the two (or more) partners. Is an integrated M.I.S. set up or do they stick to the old system?

The *primary* activities of a financial institution (whether it is a bank, an insurance company, or an investment institution) are divided into the following categories.

- Product development: we see that some financial conglomerates go beyond cross-selling each others' products. They depart from the needs of the customers and they integrate different characteristics of different products into a new product. It is interesting to see how important these new kind of products are related to the whole product portfolio.
- Market development: when insurance companies are able to use the bank's database, they can approach those clients in a better way. Banks have entry to the financial transactions of the client and have generally more information than insurance companies.
- Operations management: there are a lot of examples which prove that synergies can be achieved in this domain as well. One of the most striking examples can be found in the Eureko Group, the joint venture between some financial conglomerates in Europe (Achmea, NL - Topdanmark, DK - Ocidental, P - Friends Provident, UK - WASA, S - Gothaer, D). The Dutch partner Centraal Beheer, part of Achmea, disposed of a very good claim handling department with respect to automobile insurance. The Portuguese partner Ocidental used this system for their car insurance department. Furthermore, they installed a link with the bank branches so that the green card could be delivered immediately.
- Distribution: many conglomerates were formed because the partners saw opportunities in distributing each others' products. It is seen as one of the most important areas where synergies can be achieved. Of course there are many different ways to organise the distribution. The Dutch ING Group has organised the home market according to the main distribution channels. Other groups just cross-sell products and try to finetune the selling process.

The last category of variables are *environmental factors*. The main point of the research is oriented towards synergies and risk factors. Furtermore, integration is another important concept in the whole research project. But of course, we should analyse whether there are no other variables that influence the degree of success. For example the country in which a financial conglomerate operates could be a determining element for a firm's performance. Other important variables could be the size of the conglomerate, the time period of the partnership, etc.

It is worthwile to mention that this research project certainly has not reached its final stage. We are still refining the instrument, looking which elements are really important and which are not. However, we have tested the financial conglomerates control board on a sample of 49 Belgian and Dutch conglomerates. The results of this analysis are rather qualitative in nature and are described in chapter 5. A more quantitative analysis is given in chapter 6.

THE LEVEL OF INTEGRATION AS A REFERENCE POINT

Integration as a central element in the classification of financial conglomerates

In all the typologies we described, the level of integration is implicitly present. In the case of Herring and Santomero, the corporate structure determines the integration level although managers of a financial conglomerate may find it advisable to establish Chinese walls in order to create some operational distinction.

The typology based on the method of entry also implies that the level of integration differs according to the entry vehicle chosen. Co-operation agreements are characterised by a minimal integration, for a de novo entry it is supposed that integration is maximal. It is worthwile to mention that these are assumptions and it would be interesting to test if these hypotheses really hold. It is clear that in joint ventures and distribution alliances the two partners remain relatively autonomous. However it is not always clear whether a de novo entry is characterised by more integration than a merger or an acquisition. Empirical evidence shows us that this is not always true.

A study by Coopers and Lybrand (1993) also defines integration as a source of benefit in the financial conglomerates trend. From the results of a survey of senior executives of *bancassurance* companies they concluded that further integration depends on:

- the more opportunities and benefits financial institutions perceive in bancassurance;
- the time period the instutions are involved in bancassurance.

We refer to this study in more detail in the next chapter.

In the model we developed, integration and co-operation play a central role. As we have explained the goal of the instrument is to develop a tool to control and evaluate the drivers for and the different types of synergies within a financial conglomerate, as well as the drivers for and the different types of risk involved. It is our firm belief that synergies and the risk level are determined by the level of integration and co-operation between the different parts of the group.

Of course even if the compounding parts remain autonomous, synergies can exist in the form of *financial synergies*. A necessary condition is that income streams are uncorrelated. Only in this case, the newly combined entity may create additional value through the generation of stable income streams.

Taking a look at financial synergies

An interesting book in this respect was written by Saunders and Walter (1994). They have analysed the universal banking issue and have calculated the potential return-risk (dis)advantages of a universal banking structure based on the portfolio theory. More particularly, they have investigated whether the diversification strategies of banks into new activities (securities brokerage/underwriting, life insurance, and P/C insurance) either increase or decrease the risk of failure. They consider both advantages of risk spreading and disadvantages of engaging in more risky activities.

First they reviewed the literature on this topic and concluded from these studies:

> "The periods covered, the level of data aggregation (firm versus industry) as well as the types of methodology, data (market returns or accounting returns) and sample period differ widely across studies. Nevertheless, they generally conclude that there are some, albeit sometimes limited, risk-reduction benefits from a traditional commercial bank's diversification into nonbank activities -with most finding that securities activities are relatively risky and should be confined to a small percentage of overall bank activities, often in the 5-10% range" (Saunders and Walter, 1994: 186).

Saunders and Walter have also included different categories of insurance companies. This is a valuable contribution to the studies they reviewed because these studies looked at the diversification of commercial banks into securities activities.

The sample they used encompasses the largest financial services firms in the United States and was subdivided into the following categories:

- money center banks (9): these banks are located in the major U.S. financial centers and participate in both national and international money markets;
- regional banks (24): these banks' primary business is making

loans to customers in a regional or metropolitan area and taking deposits within the state where it is chartered;

- life insurance companies (13);
- P/C insurance companies (8): includes fire, marine, and casualty insurers;
- security brokers and dealers (11).

The risk and return data are based on daily stock returns for individual firms in the period 1984-1988. This period is characterised by important developments in the financial market (the 1987 crash, securities firms engaging in activities towards greater risk trading and the P/C insurance crisis of 1984-1985). In this way this period is characterised by major changes in the returns and risks of nonbanking activities.

This research project starts from portfolio theory and uses some general definitions of returns and risk:

$$R_u = \sum_{i=1}^{N} R_i X_i \quad (1)$$

The *return* of a universal bank (or a financial conglomerate) is the linear weighted sum from each activity it undertakes (with X_i the relative proportion of activity i in the total activity portfolio and R_i the return for the respective activity).

$$\sigma_u^2 = \sum_{i=1}^{N} X_i^2 \sigma_i^2 + \sum_{i=1}^{N} \sum_{\substack{j=1 \\ j \neq i}}^{N} X_i X_j \rho_{ij} \sigma_i \sigma_j \quad (2)$$

σ_u is the variance of returns of the universal bank, whereas σ_i and σ_j represent the variance of the returns of activity i and j. ρ_{ij} is the correlation between the returns of activity i and activity j. The *risk level* of a universal bank (defined in terms of the variance of returns) depends on the risk of each activity (the first part of the right-hand side of equation 2) but also on the correlation among the returns from the different bank and nonbank activities (the second part of the right-hand side of equation 2).

Often a composed risk measure is calculated. This *coefficient of variation of returns* (COV) is nothing more than dividing the standard deviations by the return measure:

$$COV = \frac{\sigma_u}{R_u} \quad (3)$$

This measure can be interpreted as a firm's expected return per unit of total risk. Saunders and Walter have calculated these measures for the different groups of financial institutions described before. Both C and R are measures of returns; σ and 1/COV are measures of risk. The higher this last ratio, the less risky is the industry. According to table 4.1 the fire and casualty insurance is the least risky business followed by the regional banks. Securities brokers and dealers operate in the most risky business. When it comes to profitability, regional banks and fire and casualty insurers have the best performance figures; here again the security brokers and dealers have the worst figures. It is important to notice that these figures are concerned with one particular activity.

Table 4.1 Compound returns, mean daily returns, mean daily standard deviations, and return/standard deviation ratios (period 1984-1988)

	C	R	σ	1/COV
Money Center Banks	.71732	.00061	.02024	.03014
Regional Banks	.96846	.00065	.01877	.03263
Life Insurance	.54206	.00047	.01851	.02539
Fire & Casualty insurance	1.03702	.00062	.01738	.03567
Security Brokers & Dealers	-.00394	.00027	.02541	.01063

C = compound annual 5-year returns; R = mean daily returns; σ = mean daily standard deviations; 1/COV= R/σ

Source: Saunders and Walter (1994)

If we consider the risk of the universal bank model (or in our terms the financial conglomerate model) then from equation (2) we also need correlations between the different activities. These correlations were also calculated by Saunders and Walter, both *within* and *across* sectors.

Table 4.2 Average correlation coefficients for financial firms

	Money Center Banks	Regional Banks	Life Insurance	Fire & Casualty insurance	Security Brokers & Dealers
Money Center Banks	.51722	.30275	.31142	.31326	.32276
Regional Banks		.24840	.24953	.25316	.25417
Life Insurance			.27078	.28391	.27525
Fire & Casualty insurance				.28537	.28923
Security Brokers & Dealers					.36537

The average correlation coefficients are arithmetic means of all cross-correlation coefficients.

Source: Saunders and Walter (1994)

As all these correlation coefficients are significantly different from +1, we see that profit measures are uncorrelated. This may support the arguments that there are diversification gains from universal banking (or financial conglomeration).

Saunders and Walter have also used the *merger simulation approach* to investigate whether universal banks benefit from diversification. With this approach one combines two existing firms and looks at the risk-return characteristics of this new (hypothetical) combination. For more details on this methodology we refer to this outstanding book, here we will focus on some of their conclusions.

> "The simulation analysis based on U.S. data suggests that there exists potential risk reduction gains from allowing banks to expand their activity-set in a limited fashion -and similarly for nonbank financial firms to expand their set of activities- and that these gains increase with the number of activities undertaken. The main risk-reductions gains appear to arise from banks' expanding into insurance rather than securities activities" (Saunders and Walter, 1994: 204).

Integration as measure of organisational fit

It is important to mention that the Saunders and Walter study only looks at potential risk reductions of financial conglomerates/ universal banks. In the general discussion on the diversification literature we have shown that due to inconclusive results on the effects of strategic fit, researchers have incorporated the concept of organisational fit.

And in related industries, integration is an interesting construct. Haspeslagh and Jemison (1991) have defined *integration* as:

> "an interactive and gradual process in which individuals from two organizations learn to work together and cooperate in the transfer of strategic capabilities" (p. 106).

In a study on financial conglomerates of Coopers & Lybrand (1993), integration is

> "deliberately defined in a dynamic way to reflect the *coming together* of their banking and insurance arms at any functional level" (p. 3).

Integration has been studied in the merger and acquisition literature. It has been considered as one of the most difficult stages in the acquisition process and is of particular importance to the overall acquisition performance. According to Calori and CESMA (1988), balancing centralisation and autonomy, integration and differentiation seems to be the key management challenge of diversified companies. The authors have analysed 27 large diversified companies established in France and have concluded that each company has its own formula for balancing centralisation and autonomy. Some companies opt for maximal autonomy (within certain limits), but there is always some form of strategic control, which is a form of centralisation. In this way:

"the corporate level has to give the *go-ahead* for strategic decisions, alliances are decided at the top, and reporting procedures on the strategic plan and budgets are rigorous" (p. 86).

While Calori and CESMA did not find a particular pattern on the relation centralisation-autonomy (perhaps due to their restricted sample and their methodology which was based on the case method), other researchers have developed a theoretical framework for the question: *centralisation versus autonomy?*

One of the most interesting books in this respect is a book by Haspeslagh and Jemison: *Managing acquisitions: creating value through corporate renewal* (1991). The main idea of this book is that an acquisition (or a merger) should be managed throughout the whole process in order to reap the potential benefits.

One of the most important and critical stages in the acquisition process is the integration stage. It is therefore of utmost importance to *understand the different integration approaches* which are available in the M&A process.

According to Haspeslagh and Jemison, one must focus on two central dimensions when considering a firm's integration approach:

- its relationship to the acquiring firm: this dimension relates to the nature of the interdependence that needs to be established between the firms;
- the way in which value is expected to be created: this dimension is associated with the need to preserve the acquired strategic capabilities after the acquisition.

In terms of Haspeslagh and Jemison, the *interdependence* in an acquisition depends on which kind of strategic capability transfer is the dominant source of initial value creation. That is, if resource sharing is the primary source of value creation, then it may be necessary to dissolve the boundaries of the subparts of an organisation. If the transfer of functional skills is more important, people are moved across organisational boundaries when information and know-how are shared. The transfer of general management capabilities can occur by leaving the different parts operating autonomously. Of course, new control systems and budgets may be created. Finally, if value creation originates from combination effects (e.g. increased market power) then integration may be useless.

Another determining factor for the kind of integration is the *organisational autonomy need.* Haspeslagh and Jemison argue that a lot of companies destroy the capabilities in the pursuit of capability transfer. They suggest that some form of autonomy may be necessary in order to preserve these capabilities. A good example of this paradox is provided by the financial services industry. When commercial banks have acquired investment banks, they sometimesfaced an *exodus* of the skilled personnel of the investment banks. As these investment bankers are the main source of value creation, the acquisition may become worthless when these people decide to leave.

In order to deal with this paradox, Haspeslagh and Jemison suggest that autonomy should be provided if the survival of the strategic capabilities on which the acquisition is based depends on the preservation of the organisational culture from which they came.

Based on these considerations, Haspeslagh and Jemison have developed a typology for different kinds of integration approaches. This typology is presented in figure 4.3.

Figure 4.3. Types of integration approaches (Haspeslagh and Jemison, 1991)

Need for organisational autonomy	**Need for strategic interdependence** Low	High
High	Preservation	Symbiosis
Low	(Holding)	Absorption

Here follows a short description of the different integration approaches.

- *Absorption*: here the objective is to dissolve the boundaries between the two units.
- *Preservation*: as we can deduct from figure 4.3, the need for strategic interdependence is low, while the need for organisational autonomy is high. Value is either created by nurturing (e.g. the main purpose is to bring in the necessary funds -some kind of financial synergy) or by learning.
- *Symbiosis*: according to the authors this kind of acquisition is extremely difficult to implement because high strategic interdependence is needed, while on the other hand the parties have to remain autonomous.

> "In symbiotic acquisitions the two organizations first coexist and then gradually become increasingly interdependent. This coexistence and mutual dependency are slowly achieved despite the tension arising from the conflicting needs for strategic capability transfer and the maintenance of each organization's autonomy and culture" (p. 149).

Based on this typology, Haspeslagh and Jemison define key integration challenges for the different types of acquisitions. It is of great importance to determine the type of acquisition because integration issues have to be solved in different ways. For example, in absorption acquisitions managers have a clear-defined plan which should be executed as fast as possible. However, in preservation acquisitions this process is more evolutionary while for symbiotic acquisitions, an adaptive attitude is more appropriate.

Other researchers, although a minority, have considered the integration aspect of a merger or an acquisition. Shrivastava (1986) has distinguished between procedural integration, physical integration (such as product lines, production systems), and managerial and sociocultural integration. But here again, the author warns for the appropriate level of integration: underintegration may be unproductive, overintegration may be expensive. Much depends on the motives of the merger: e.g. if the purpose of the merger is to exploit multiple synergies, a comprehensively integration is necessary; if on the other hand the purpose is to increase market share in a limited product/market domain, partial integration is sufficient.

The concepts of centralisation and autonomy highlighted

In the description of Haspeslagh and Jemison's typology, two concepts were central: interdependence and autonomy. Other researchers have focused on these concepts and on their interrelations. While interdependence and autonomy originally came from merger and acquisition literature, soon the concept of diversification was introduced into these discussions. For the sake of clarity, we first define a *decentralised* firm as a firm where the subordinate manager is responsible for the financial performance, including the profitability, of "his" or "her" business (Vancil, 1979: 1). As a matter of simplification we consider decentralisation and autonomy as synonyms, although Vancil sees some important differences[4].

Datta and Grant (1990) consider the type of acquisition (related vs. unrelated) as an important contingency factor which determines the degree of autonomy (decentralisation) and centralisation. Results show that in unrelated acquisitions the extent of autonomy is significantly greater than in related acquisition. Also performance is positively associated with superior performance in unrelated acquisitions. Similar results were found by Davis e.a. (1992) who investigated the paper and pulp industry: relatedness (measured in terms of functional interdependence on marketing and production

---

4 Vancil defines *autonomy* as the manager's perception of his power to initiate change and his ability to influence events (Vancil, 1979: 51). Of course a manager has only real power if he has the associated *authority*, i.e. the disposal of physical resources. Vancil has described this relation as follows:
"A profit center manager who has substantial functional authority also believes that his *autonomy* to take independent action is relatively high. This relationship is not surprising, of course, because both indices are measures of the independence of a profit center manager; one measure is physical and the other is perceptual. The strenght of the relationship, given the crude nature of our authority index in particular, *is* surprising" (p. 70).

level) and autonomy were found to be inversely and significantly related. This is quite important because decentralisation and granting increased managerial discretion to operating management is generally seen as desirable. However, increasing interdependence between different business units places constraints on this autonomy, and thus may be a cost which must be taken into consideration. This was also suggested by Gupta and Govindarajan (1986):

> "at the level of individual SBUs, there are costs of reduced flexibility, because sharing resources reduces SBU's managers' independence with respect to the allocation of those resources and reduces their flexibility in responding to unanticipated competitive moves."

It seems that exploiting the synergies is an exercise which requires an optimal balancing between achieving interdependencies between the compounding parts and avoiding the problems of co-ordination and control loss.

Or in the terms of Porter (1985):

> "uncontrolled decentralization, however, undermines the pursuit of interrelationships by encouraging business unit managers to pursue strategies that maximize unit performance and not the corporations... The organizational difficulties of achieving even clearly beneficial interrelationships is perhaps the single biggest reason many managers have rejected the concept of synergy" (p. 383-384).

PART B

Testing the economic value of all finance and financial conglomerates

5

COMPARATIVE ANALYSIS OF DIVERSIFICATION STRATEGIES WITH THE HELP OF THE FINANCIAL CONGLOMERATES CONTROL BOARD

INTRODUCTION

This chapter describes how the *financial conglomerates control board* can be used to classify different financial conglomerates. This research project is rather qualitative by its nature, and the purpose is to situate different financial conglomerates by the different aspects of this instrument.

In the previous chapter we have argued that this instrument is an appropriate means to determine the *drivers of synergy* (and the drivers of risk).

More particularly, with this instrument we want to depict where different financial conglomerates try to achieve interrelations between the banking and insurance business. As it are these interrelations that (are supposed to) create added value (and thus justify the existence of these new financial institutions), we want to analyse the main drivers for synergy and how synergy is given shape.

On the other hand, combining different activities may create *additional risks* which may not be present if both businesses were separated. We refer to these risks in chapter 10, but it is worthwhile to point to an interesting aspect in this respect. A lot of these risks are concerned with the relation between the financial conglomerate and the supervisor. The new structures which are formed are more difficult to supervise and therefore there is potential risk of abuse by the financial conglomerate due to information asymmetries. As such only a few risks apply to operational risks faced by financial conglomerates. However, our main focus is on the operational aspects (e.g. operational synergies) concerning these interrelations between banking and insurance businesses. That is also the reason why we skip these risks for the moment and discuss them later on.

We have identified 25 Belgian and Dutch financial conglomerates for which we analysed the different elements of the financial conglomerates control board. We have defined financial conglomerates in the strict sense: it is essential that there are capital links between the banking and the insurance partner. We also included joint ventures where both partners have a qualifying holding[5] into a newly created company. We did not consider financial groups where the only link between the banks and the insurance companies was based on a co-operation agreement. We have to admit that a lot of financial institutions have those kinds of agreements trying to benefit from cross-selling opportunities. However as these co-operation agreements are restricted to distribution (and sometimes marketing) activities, these firms have not many options for further integration.

We have opted to study Belgian and Dutch companies because the information we needed was quite detailed and could not always be retrieved from the annual reports. We relied on articles of financial newspapers (*Financieel Economische Tijd* for Belgium, and *Financieel Dagblad* for the Netherlands) and other articles from specialised financial magazines and other sources.

We collected information from the following financial conglomerates.

Table 5.1 Overview of the financial conglomerates under study

ABN-AMRO (NL)	Generale Bank (B)
Achmea (NL)	ING (NL)
Aegon (NL)	Landbouwkrediet-Swiss Life (B)
AGF/Assubel (B)	Nuts-Ohra (NL)
Almanij-Kredietbank Group (B)	P&V-Bank Nagelmackers (B)
Argenta (B)	Rabobank-Interpolis (NL)
Assuranties Groep Josi (B)	Reaal Groep (NL)
BACOB-DVV (B)	Royale Belge-Ippa Bank (B)
BBL (B)	SNS-NOG (NL)
CERA-ABB (B)	Stad Rotterdam (NL)
Delta Lloyd (NL)	De Vaderlandsche (B)
Fortis (B-NL)	Winterthur-BBL (B)
Gemeentekrediet-OMOB (B)	

(B): Belgian - (NL): Dutch

As already mentioned our financial conglomerates control board consists of three parts. In this chapter we will focus on (1) the structural elements and (2) on the integration of the different activities.

[5] We defined qualifying holdings as equity stakes in which the company has the possibility to influence the management of the other company to some extent.

LOOKING AT THE STRUCTURAL ASPECT OF THE DIFFERENT FINANCIAL CONGLOMERATES

The legal structure

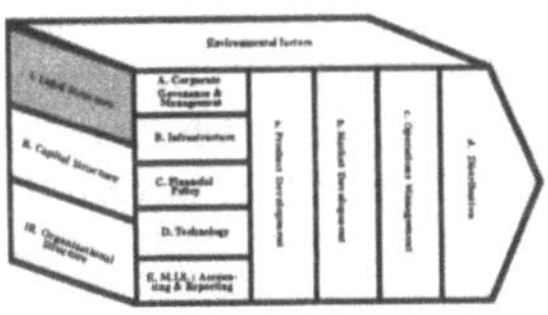

When considering the legal structure of a financial conglomerate, we primarily observe how the financial conglomerate emerged. Different *methods of entry* are possible: start-ups, mergers and acquisitions, joint ventures, and alliances. This is indeed the classification which was used in the earlier studies on bancassurance and all finance.

A number of interesting elements can be mentioned in this respect. First, within a financial conglomerate *different routes* are possible. One of the most notable examples is the Fortis Group. This Belgian-Dutch financial conglomerate is the result of a merger between two financial conglomerates itself. The Dutch part is composed by one of the largest Dutch insurers AMEV which merged with VSB, the largest savings bank group in the Netherlands. This group was formed in January 1990 and was broadened considerably by the merger in May 1990 with the Belgian insurance group AG. AG disposed of a small bank -Metropolitan Bank- but had established close co-operation agreements with Belgium's leading bank, Generale Bank (Générale de Banque). AG and Generale Bank had set up a joint venture, Alpha Life, to sell life insurance via the branches of Generale Bank's network. The alliance with Generale Bank ended when Fortis decided to take over the ASLK (CGER) Group. This state-owned insurance and savings bank group was the first bancassurer in Belgium as the relation between the ASLK Bank and ASLK Verzekeringen dates back to the 19th century. Even then the insurer made already use of the bank branches of the bank. From that moment on Generale Bank acquired all shares of its joint venture Alpha Life, and AG concentrated on its Fortis activities. This example proves that there exist large groups, consisting of different clusters, each with their own characteristics and peculiarities (e.g. on the level of distribution). Thus within a group different legal structures are possible.

Furthermore it is important to focus on the *dynamic* aspect of legal structures. This is especially the case for distribution and marketing alliances and joint ventures. It occurs frequently that after a certain time two (or more) partners end up with tighter relations. Frequently joint ventures led to a sale by the insurer to the bank. As banks gained experience in selling and underwriting risks, they considered the insurers somewhat worthless after a certain time. In Bleeke and Ernst's typology (1995) this kind of joint venture is called *evolution to a sale*. According to the authors, these alliances start with two strong and compatible partners, but competitive tensions and bargaining power may shift so that one partner ultimately sells out the other. Then the newly created (life insurance) company becomes a subsidiary of the bank.

There are some remarkable differences between the evolution in Belgium and the Netherlands. It seems that in Belgium the evolution towards tightly integrated financial conglomerates is much slower than in the Netherlands. It is remarkable that before the all finance and financial conglomerate development really boomed, there were yet a number of alliances and capital relations between banks and insurance in Belgium. In the Netherlands this occurred less frequently. There the liberalisation of the Structure Policy in 1990 really changed the financial services industry dramatically. As we will see, the Dutch financial conglomerates also aim at a more radical renewal of their strategies, while in Belgium the adaptation to the new competitive environment occurred more moderately.

The capital structure

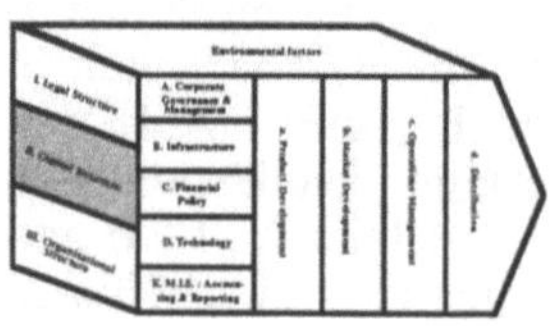

We looked at financial participations within the group in order to observe how financially related the group is.

We expect that tight operational relations are expressed by tight capital relations. This must not necessarily equal 100 per cent, but often the capital stakes exceed 90 per cent. On the other hand when two firms decide to create a joint venture the parenting companies generally have small capital participations into each other.

However these general observations do not always hold. Not all groups aim at large equity stakes in their subsidiaries. This is predominantly the case for some Belgian groups which have less pronounced equity relations, although they have control over their subsidiaries. This is the case for the Belgian Almanij-Kredietbank Group. Almanij acts as the holding company of a group which consists of several banks -Kredietbank, Crédit Général, HSA, Spaarkrediet, Bank van Roeselare and KB Lux- and other specialised financial institutions (insurance companies, brokers, leasing companies, etc.). According to Jan Huyghebaert, chairman of the board of Almanij, the main idea is that Almanij wants to govern the relations between these financial institutions but leaves the initiative to the subsidiaries. This is also expressed in the way the subsidiaries are consolidated: Almanij uses the equity method.

Here again there are remarkable differences between financial conglomerates in Belgium and the Netherlands. Belgian financial conglomerates are sometimes characterised by very complicated group structures. A good example in this respect is the structure of the Argenta Group, a relatively small but very integrated financial services provider (see figure 5.1). These relations are set up in this way in order to stimulate co-operation between the respective business units in the group.

Figure 5.1 Group structure of Argenta (Annual report, 1995)

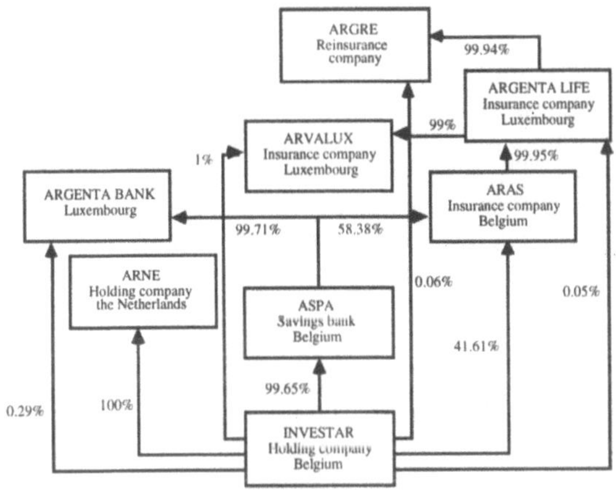

In the Netherlands a considerable number of financial institutions are mutuals which are owned by their policy or account holders. These firms also have a different viewpoint on the importance of capital structures: governmental influence is much more important than the question how the shares of the subsidiaries are distributed. This statement was confirmed by a manager of the Rabobank-Interpolis Group who stated that equality in co-operative terms is something different than counting guilders.

Perhaps this is an indication that capital structure per se is not sufficient to investigate the integration of financial conglomerates.

The organisational structure

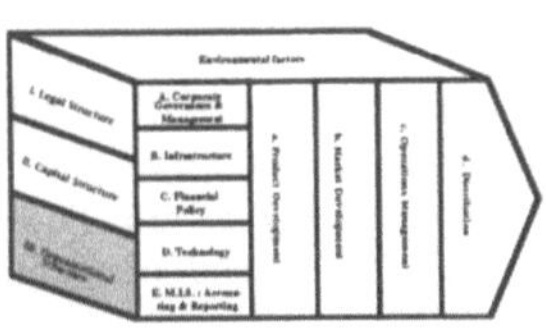

Financial conglomerates differ in the way they implement their diversification strategies. According to Chandler structure follows strategy, and thus there should be differences in the organisational structure of financial conglomerates. And indeed this is the case.

Broadly speaking we can distinguish four organisation structures when two (or more) firms decide to form a financial conglomerate.

- The first option is to maintain the existing structures of each partner as they are. This means that the banking and insurance businesses operate relatively separately and no ties exist at the structural level. This option is taken when the firms have chosen a loose form of co-operation, aiming at autonomy of the respective businesses. Sometimes there are links between the partners on the management level (see later).
- The second option is to create a holding company for the different business units. In this way the business units operate autonomously but are controlled by a central head office. There may be some activities such as treasury, risk management, investment which are dealt with centrally. To a certain extent this may lead to the absorption of one company by another. This integration can be chosen when two business units serve the same market with the same type of product.
- The third way financial conglomerates have organised their activities is by creating a holding company as well as joint and administrative centres. These centres are an additional layer between the head office and the business units. This method is used when the group is large and consists of several business units. Furthermore, financial conglomerates which aim at a strong integration have opted to create these additional layers because this subdivision is more relevant for their operational strategy.
- The last option occurs when the legal structure is a start-up or when a large company acquires a small one. In this case the smaller company is the subsidiary whose activities are in function of the strategy of the parent company. No complicated organisational structures are set up because the co-operation involves only a few companies.

These four options are presented in the figure 5.2. It is clear that the most drastic changes occur in the large financial conglomerates, aiming at strong integration of their banking and insurance businesses (situation 3). These financial conglomerates (such as ING and Fortis, but Argenta as well) have set up organisational structures which go beyond the traditional boundaries of the financial services industry. The administrative layers are organised according to the distribution channel used and the target client segments.

Figure 5.2 Schematic representation of the different organisation structures

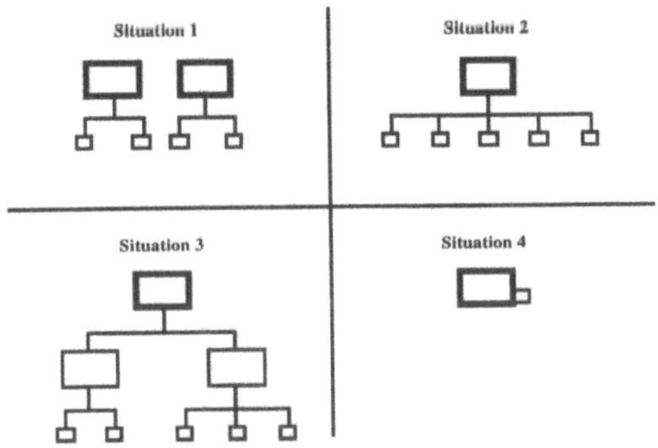

Of course for legal purposes a strict separation between banking and insurance companies is necessary. And therefore there may exist two structures: a structure which is suited for supervisory and other legal requirements, and a structure which is the real base for the strategic decisions.

INTEGRATION OF THE DIFFERENT ACTIVITIES OF FINANCIAL CONGLOMERATES

In this section we investigate whether the financial conglomerate aims at the integration of the different activities involved. We look whether there is integration between banking and insurance activities. In general this integration can vary from centralisation (maximal level of integration) to minimal co-operation and co-ordination between business units.

Corporate governance and management

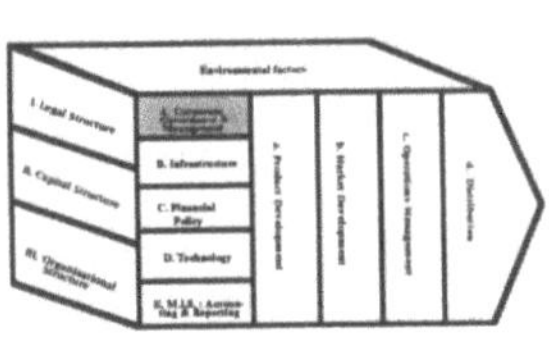

The *corporate governance and management* level consists of different activities: the governance and management itself, including also strategy development, human resource management, and initiatives to manage the different cultures. For each of these activities different integration strategies are possible. However we classified them under the same topic because these activities all have in common that they are involved with the way the strategy is implemented.

Here again we see large differences between the various financial conglomerates. On the one hand, some financial conglomerates opt for tight management structures and strive for an integrated group approach. Generally we can state that an integrated management team has *strategic control* over the business units, which in turn have operating autonomy. In order to create the necessary flexibility these groups make use of a quite integrated human resource management. Mostly the personnel department is centralised. Furthermore these groups make use of job rotation and cross-overs, i.e. techniques whereby managers and other personnel are rotated through the whole company, once in a banking subsidiary, once in an insurance company. Sometimes these companies have created collective agreements per industry for all companies of the group (within a particular country). Attention is paid to prevent culture clashes and to stimulate synergies. Examples of such financial groups are Fortis, ING, SNS-NOG, Reaal, Argenta and Achmea.

When one company has *absolute power* over another company (e.g. large bank disposes of a small insurance company or vice versa -see situation 4), this is also reflected in the composition of the board of directors and in the strategy formulation. Generally managers are controlled by a board where the majority are managers of the parent company. The strategy of the subsidiary is developed in function of the strategy of the parent company, and e.g. the same client typology is used and targeted for. Sometimes the personnel of the smaller company has the same labour conditions but this is not always the case.

If the partners remain autonomous, minimal links on the management level may exist. For example there may be a representation in each other's board, but the managers of the compounding parts act independently. Strategy formulation is decentral and there is almost no integration of the human resource department. This is the case when partners in a group have considered the benefits of synergy only recently (although the group may exist for decades). Examples in this respect are the combinations CERA-ABB and BACOB-DVV. At the same time it can be stated that these too make increasing efforts for further integration. Of course this pattern can especially be found in companies which have rather loose legal structures such as joint ventures, and marketing agreements.

Infrastructure

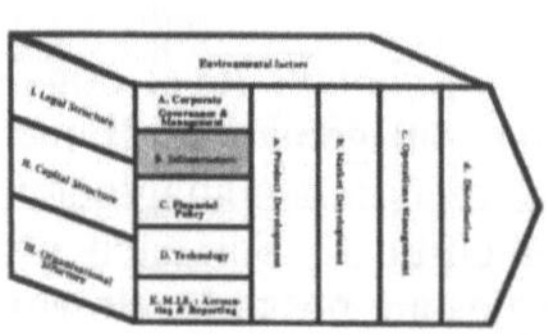

From the interviews we held with managers of financial conglomerates, we understood that integration of the infrastructure (such as offices and systems) does not occur very often. Only some small groups have chosen to integrate their banking and insurance subsidiaries. But this is certainly not the rule, rather the exception.

The larger groups sometimes regroup some business units and centralise some companies but the business units operate from distinct locations. When there is one management team for the whole group, then there is generally a central head office. But according to a manager of a very integrated financial group: integration of the banks and the insurance companies on the level of infrastructure is absolutely not necessary.

Financial policy

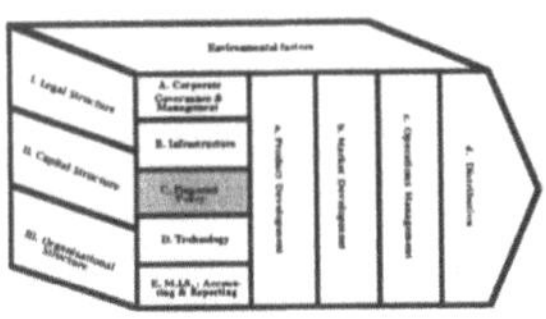

In the preceding chapters we argued that the outflow of their deposit base forced the banks to set up their own (life) insurance companies. In this way the banks were able to recapture the money they had lost to life insurers. Consequently, a financial reason was one of the main motives for setting up a financial conglomerate structure.

Indeed the investment function is very important for both insurance companies and banks. The investment income of insurance companies determines their profitability to a large extent. In the life insurance industry the results of the investments are an indicator of company performance. The investment function of a bank is perhaps even more important because it lies at the heart of the core business of a bank: the transformation of money. Therefore, if considerable synergies can be achieved on the investment level, integration of the investment department may be a critical success factor of financial conglomerates. Economies can be realised by means of a reduction of search and transaction costs as well as on the level of improved know how and learning experiences.

We notice that here again the strategies of the financial conglomerates differ considerably. The more integrated companies opt for a centralised investment department. Sometimes this is formalised in separate organisation structures where the investment function is considered to be a separate business unit. We refer to the strategies of the large integrated financial conglomerates such as ING and Fortis. The conglomerates with a bigger and smaller partner also integrate the investment function but the co-operation is not always formalised.

However, some financial conglomerates which integrate only a limited number of activities. For example insurers make use of the buy and sell capabilities of their banking partners when offering investment funds. Bernard De Gryse, managing director of the insurance companies of BBL, sees no opportunities for more synergies, because the investment activity of insurance companies is totally different and requires a different Asset Liability Management.

The financial conglomerates with rather loose co-operation structures have just discovered the wide spectrum of potential synergies on the investment level and co-operation initiatives are rather an exception to the rule. However, we see that this is changing. More and more both banking and insurance partners have found each other and start developing initiatives in this respect. But most of the time these initiatives are relatively modest.

(Information) technology

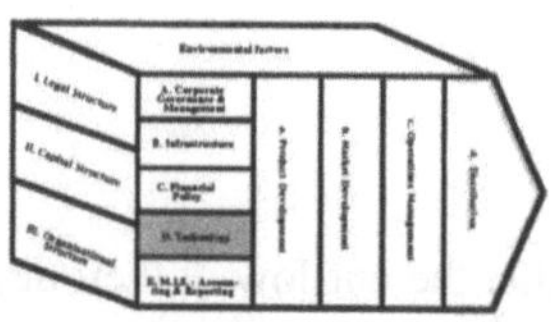

It was more difficult to evaluate the synergies on technology because synergies in (information) technologies can occur in many different ways.

We found two examples where the use of IT was really a source of process improvement. The most cited example was the use of an integrated front-office system. In most banks there is an electronic acceptance system so that the personnel of the bank branch can handle the insurance contract. In this way there is only one input procedure which decreases potential errors. In some cases there is a direct link with the insurer's back office so that the necessary documents can be printed in the bank's outlet. There are examples where the green card of the car insurance rolls out of the bank's printer a few minutes after the insurance contract was signed.

Another example is the electronic handling of securities. We have already mentioned that at the investment level banks and insurance companies work together for the administration of securities. Banks have a range of funds to which the insurance product can be directly linked (unit linked products) or indirectly invested (investment of the technical reserves). As more and more consumers are interested in these kind of savings vehicles the electronic handling of these securities may rise in importance, and thus IT may be an important means to control efficiency.

It is often mentioned that both partners can have access to each other's customer base. In this way a central database for the whole group may be created so that each business unit has the possibility to ask information of a particular customer. It is clear that information technology may play an important role in this process. To our knowledge one financial conglomerate is working on such a database warehouse project in which both the insurance and the banking subsidiaries have access to client information. Of course there are legal restrictions for what concerns privacy, but financial institutions try to link the different databases of their subsidiaries.

Of course IT may be used for other activities such as planning and control, marketing, payment activities etc. There is however no clear tendency how IT is involved in all these processes. One thing is certain: the importance of IT and technology in general is rapidly increasing.

M.I.S.: Accounting and reporting

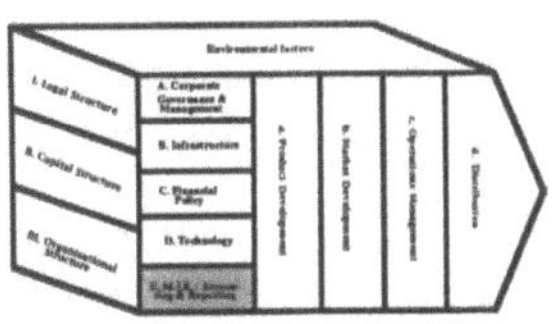

As was the case for the technology function, here again it is quite difficult to detect the relative integration efforts between the different financial conglomerates. If we face some large groups, whether they are integrated or not, there is a need for controlling the subsidiary and therefore a management information system must be developed.

In the general literature a distinction is made between pure financial controls and more strategic controls. We think this distinction is also applicable to the financial services industry. If financial conglomerates aim at a strong integration of different business units, then attention must be paid to strategic control. If on the other hand banks see their insurance companies as an investment (or vice versa), financial controls may be more appropriate.

However, even in strongly integrated financial conglomerates the main focus is on a number of financial criteria. Most of the time the performance of a business unit is benchmarked to some fixed financial measure (such as return on investment). Furthermore if there is a particular focus on more operational aspects, the business units are free to determine how they will fulfill the requirements laid down in the strategy process.

Of course, in the situation where a larger company has control over a smaller company, the accounting and reporting function may be more integrated, because the activities of the smaller company are designed in function of the larger parent company. Also in larger integrated financial conglomerates there is a clear tendency towards more tuning of the M.I.S.

Product development

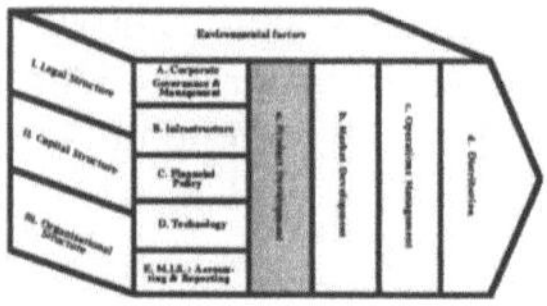

When a financial conglomerate aims at integration at the distribution level (see later) there is always some form of integration in the product development phase, because the products have to be fine-tuned to the distribution channel that will be used in selling the product to the customer.

For a long time all business units had their own product development department. It was thought that the operating activities, such as product development, were to be carried out by the business units themselves and no centralisation was necessary. But as the groups become larger, and one sees that some applications invented by a certain business unit can also be used by other business units throughout the group, there is more and more communication and a need for co-operation. This development is further strengthened by the fact that some new financial products are combinations of existing insurance or financial products. These developments force the bank and the insurance company to work together.

The main idea for the future is to work with modular concepts. These modules form the building blocks around which new financial products will be built. This new approach allows for some standardisation (there are only a few modules) but on the other hand the combination of different modules and other features makes it possible for a financial firm to develop products tailored to the specific needs of a customer. For the moment this is best noticed in the planning phase: a small number of companies are thinking of applying this idea in practice shortly.

Market development

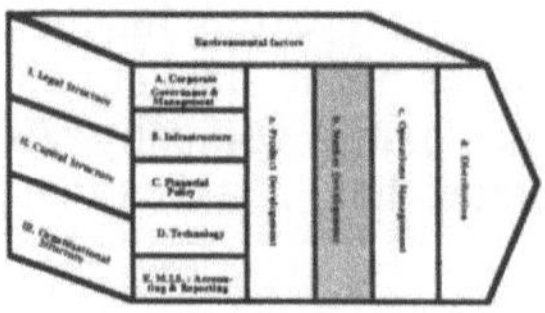

Market development is a term invented by Ansoff (1965) to indicate a growth strategy by which a firm tries to develop new markets with existing products. We attach a similar interpretation to this concept: we want to look how the different partners develop a market for products of their partner. In the literature on bancassurance, it is often stated that personnel of the bank branch provides insurance sellers with *warm leads*. In this way the branch staff is expected to prepare the sale and depending on the nature of the product either the bank's personnel or an insurance representative handles the sale. But market research in Britain has shown that the relationship between banks and customers is not always that warm.

> "Over half the respondents said that no one at their bank or building society knew their name and almost two thirds of respondents did not know the name of any of the staff that worked there" (Money Management, December 1994: 59).

However, it is quite difficult to develop measures of integration of market development on a corporate level. This is the case because the process we described here is rather at work on the business unit level.

If we look how financial conglomerates give shape to their *corporate communication*, this may shed some light on how integrated their market development is. The corporate communication function changed enormously during the last decade. Large groups have emerged and the appropriate question is whether to use a uniform brand name for the whole group (uniformity model) or to allow different brand names within the group (variety model). In case the latter model is preferred, we can raise the question how much space is left for a distinct profiling of the different business units. There are no clear answers to these questions.

The large financial conglomerates with substantial banking and insurance subsidiaries seem to opt for the variety model. But they also tend to create a corporate image. On the other hand if the number of business units is restricted or if there are large differences in size, financial institutions are more inclined to change the name of the smaller subsidiaries into the corporate name. Examples are: OHRA, BBL, Delta Lloyd, De Vaderlandsche, ABN-AMRO and Stad Rotterdam. When the banking and insurance groups remain relatively autonomous there is no integration of the corporate communication function at all.

It is interesting to refer to a study of van Riel (1995) who argued that the choice between a uniformity model and a variety model depends on several factors.

- Strategic fit: if there is complementarity in the scope of the different activities and similarity of market approach, a uniformity approach may be more appropriate. If a financial conglomerate consists of two business units in which firm A operates very aggressively and firm B acts more conform to the average market behaviour, the financial conglomerate may choose the variety model.
- Organisational fit: if there is a tendency to centralise the control and planning function, the corporate communication

may also be centralised so that a more uniform approach is in place.

- Driving forces fit: if the employees identify themselves primarily with the group a uniformity approach is more appropriate, on the other hand if there exist strong business unit cultures it is better to work with several brand names.
- Environmental fit: it is clear that if a brand name is well-known by its target public the tendency to replace it with a less-known brand name is smaller.

Operations management

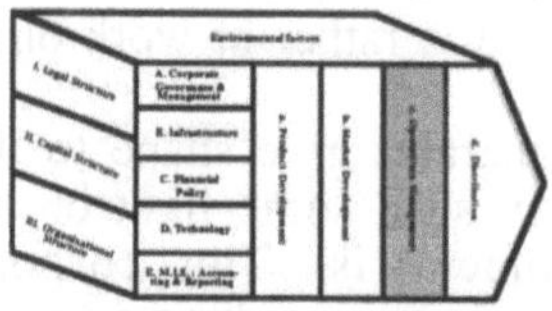

There are no clear patterns regarding integration of operations management. This question is much more a matter at the business unit level so no generalisations on the corporate level can be made.

However what we perceive is that all financial conglomerates try to integrate production and selling processes (see also point 5.3.4 on information technology). There is not only a tendency to integrate the administration of the selling process but also to incorporate the acceptance process of insurance products in the bank. Thus the bank's staff has the authority to select the insurance risks themselves.

Distribution

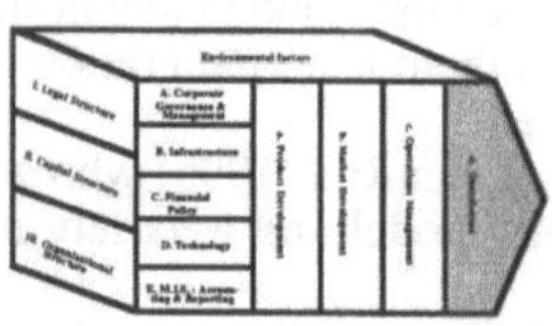

Most financial conglomerates primarily aim at distribution synergies. Banks and insurance companies agree to cross-sell each other's products and thus benefit because an additional distribution channel is used. However this is not a general strategy because some groups have opted to serve only one distribution channel. For example groups such as Delta Lloyd, Stad Rotterdam and De Vaderlandsche have strong ties with independent insurance brokers. They use their banking subsidiary to deliver appropriate banking products (mostly investment products) to this distribution channel.

As these benefits may seem straightforward, we have already indicated that severe problems can arise. If an insurer or a bank tries to provide additional distribution channels with their products the existing distribution channels can consider this as an attack on their positions. This explains why some groups have reached distribution agreements for a wider range of products only after a few years. This is certainly the case for companies that work with (in)dependent brokers and agents but also banks can stick to this strategy. For example, Rabobank has one of the most intricate distribution networks in the Netherlands and these branches are consciously used to give shape to its all finance strategy. However Rabobank avoids appealing to the insurance intermediaries for selling banking products because this could be at the expense of the existing branches.

Furthermore it is very difficult to determine the appropriate commission structures for each distribution channel. Thus, exploiting the distribution synergies has much to do with balancing on a tight-rope. However, it seems that multichannel distribution has become an accepted fact in the financial services industry and that each distribution channel has to position itself in this new competitive environment.

Illustrative for this new approach is the answer of Ewald Kist to the question how ING succeeds to be in good terms with all distribution channels:

> "We don't sell ING insurance products. Each distribution channel has own brand names (Nationale-Nederlanden, RVS, Postbank Verzekeringen) and own products. Insurance products designed for the Postbank are totally different, have an own price and are standard products. Insurance policies sold by the independent brokers are tailored products which require professional advice and also have an own price" (De Financieel-Economische Tijd, Oct. 22th, 1997).

As distribution synergies are considered to be very important, more and more financial conglomerates have set up organisation structures based on their multichannel strategy. It seems that splitting up the concern according to the different distribution channels used is the most appropriate way to segment the market.

THE FINDINGS OF THE COOPERS & LYBRAND STUDY: MAKING BANCASSURANCE WORK

In the preceding paragraphs we have indicated where and how financial conglomerates have tried to achieve synergies. We defended that the source of these benefits is the integration of structures and activities. Before we empirically test whether more integration leads to better results (see chapter 6), we would like to focus on a study made by Coopers and Lybrand: *Making bancassurance work* (1993). This study was based on in-depth interviews with senior executives in the head offices of over 50 banks and insurance companies in the major European countries.

Coopers and Lybrand have come to similar conclusions regarding the importance of integration:

> "Despite differences in priorities and approaches for the implementation of bancassurance[6], over half of the institutions interviewed indicate that they are moving towards fully integrated approaches to sales, distribution and marketing support, whilst a significant minority also intends to integrate a number of other business functions through bancassurance" (Coopers & Lybrand, 1993: 3).

Furthermore the study revealed a positive correlation between:

- the level of integration and the range of recognised benefits of bancassurance, and
- the level of integration and the time financial institutions were involved in bancassurance. It seems that experience breeds awareness, which in turn leads to further integration.

Another interesting element is that Coopers & Lybrand found differences with regard to the all finance strategies of banks and insurance companies. Banks plan to integrate further than insurance companies, at least their sales forces and distribution channels. On the other hand insurance companies plan to achieve more economies of scale. These somewhat opposing results just indicate that banks and insurance companies have different purposes for bancassurance and all finance: one of the main marketing advantages for insurance companies is that they can make use of the customer database of the banks. This is an advantage which cannot be exploited by banks because insurers cannot easily reach their client base directly if they work with independent brokers.

The results also show that insurance companies also expect more benefits from using banks for improving contacts with financial markets (p. 14). Stated differently, insurance companies expect that they can benefit more from banks for the investment function than vice versa. In this way the insurers hope to learn the financial techniques the banks use.

One of the main recommendations of *Making bancassurance work* is to move towards further integration. According to Coopers & Lybrand the ideal path is to follow the value chain backwards. One should start integrating sales and marketing activities, and afterwards integration should be expanded to other business functions. According to Coopers & Lybrand this is necessary because the intention of the vast majority of those involved in bancassurance to move towards more integration is driven partly by the desire to meet customer needs more effectively. As the sales and marketing functions are the main contact points with the client, it is obvious that an optimal fit between banking and insurance business units is necessary to meet the needs and wishes of the customer.

---

[6] In this context bancassurance is used as a synonym for assurfinance and all finance.

Figure 5.3 Integration should follow the value chain backwards... (Coopers & Lybrand, 1993)

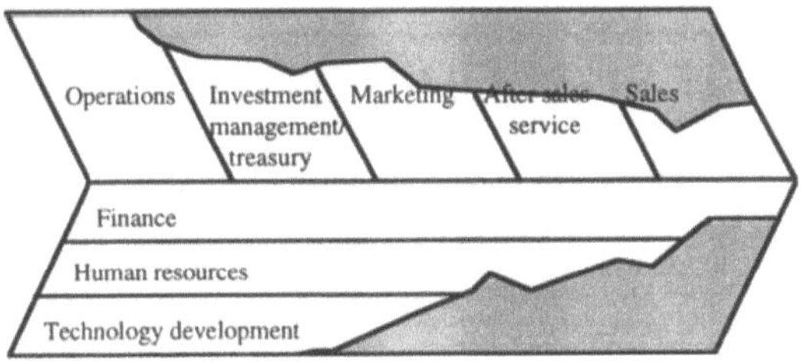

According to the findings of our analysis based on the financial conglomerates control board, this statement is not completely true. It is true that experience breeds awareness and that this determines further integration. But we think that as long as there is no integration at the top as well, the opportunities and benefits remain restricted. There are just too many difficulties and problems in adequately implementing an all finance strategy. These problems require some form of central co-ordination and management. A similar vision at the top level helps to remove the hurdles that have to be overcome in order to successfully implement an all finance strategy.

6

RESEARCH INTO THE OVERALL EFFECT OF THE DIVERSIFICATION STRATEGIES IN THE FINANCIAL SERVICES INDUSTRY

INTRODUCTION

The purpose of this chapter is to analyse the performance effects of the diversification strategies within the financial sector. Diversification and more particularly the relation between diversification and performance, has been well-documented in strategic management, finance and to a lesser extent in industrial economics literature (Hoskisson and Hitt, 1990; Ramanujam and Varadarajan, 1989). Despite the great volume of research in this area performance effects of firm diversification remain undecided. This is due to different performance measures, diversification measures, samples and time periods (Lloyd and Jahera Jr., 1994). Johnson and Thomas (1987) are not surprised with the different findings: the findings are likely to differ because the contexts differ. Their solution to this problem is to analyse diversification within a specific industry. This is also the approach we have adopted.

> "In many ways the results on diversification strategies differ markedly from those of the stream of across-industry studies which focus on large firms and largely ignore industry effects" (1987: 359).

This contextual approach is also defended by other researchers (Hill and Hansen, 1991; Datta e.a., 1991; Brush, 1996; Dess e.a., 1995) who state that:

> "the causes and consequences of diversification are industry-dependent... By examining diversification independent of its industry context prior studies may have glossed over the complexities of the diversification-performance relationship" (Hill and Hansen, 1991: 187).

Diversification literature also suggests that (future) research should include different measures of diversification. Moreover, strategic researchers should also focus on *strategy implementation* when considering the diversification-performance relationship (Hoskisson and Hitt, 1990; Nayyar, 1992). One of the key elements of successful diversification is the managing of post-acquisition integration (Haspeslagh and Jemison, 1991). We have developed an industry-specific instrument which measures the degree of integration within a financial conglomerate and we will test *whether integration is a necessary condition for successful diversification.*

First we give an overview of the most important developments in the European financial services industry and more particularly we will focus on the Belgian and Dutch situation. This will not only yield better insight into the relevant key strategic dimensions of this industry but it will also allow us to make more precise statements about the diversification strategies of financial firms. For a review of the theoretical concepts, we refer to chapter 2 (more particular paragraph 3 where the concept of *synergies* is discussed in detail). Based on previous research findings, we formulate the hypotheses which are empirically tested on a sample of 49 Belgian and Dutch financial groups. We conclude this chapter by evaluating the results and by indicating some suggestions for future research.

Diversification in the European Financial Sector: Focus on the Dutch and the Belgian Situation

In the preceding chapters we have extensively described the motives and the diversification process. We have argued that the diversification strategies have reformed the financial services industry thoroughly. We also pointed out the huge heterogeneity in the further development of this trend: a wide range of organisational structures has been developed to form these financial conglomerates. Some of these financial conglomerates have opted for rather tight relationships and have merged, acquired or set up an insurer or bank. Others have chosen looser forms of co-operation such as joint ventures and marketing agreements. However, the first publications which described the diversification trend in the financial sector pointed out that in order to achieve success some form of integration was necessary. More integration should result in a better exploitation of synergies, and this in turn should lead to higher performance. Not all financial conglomerates have followed this route of thinking and have kept the banking and the insurance part separately. The main reason for this 'split up' is that they consider banking and insuring to be two distinct professions, and the cultural differences just too large to integrate these businesses successfully.

It is important to notice that there remain specialists as well. These financial institutions reject the idea that a combination of banking and insurance activities is beneficial and influences performance in a positive sense. As a conclusion, we can state that *diversification* (both content and process aspects) is an important characteristic that differentiates the firms in this industry. However, other characteristics may determine performance of financial institutions as well.

Not only have there been a lot of takeovers between banks and insurance companies, the merger and acquisition activity within the particular sectors was also very high. Where diversification strategies primarily aim at economies of scope, horizontal mergers have the purpose of achieving economies of scale. E.g. According to Vander Vennet (1996) the increase in EU bank takeovers is generally ascribed to the *1992-effect*, i.e. the effect of the European integration. It is argued that mergers and acquisitions will provide banks with opportunities for efficiency improvement. Despite a huge stream of (predominantly American) research indicating that economies of scale were restricted to rather small levels of output, banks acted like there were a lot of economies to be gained. Furthermore, the banking market is growing (e.g. the introduction of the *Euro* will create a European market) and the current players use mergers and acquisitions in order to maintain their relative position. The same story holds for the insurance industry: in order to gain critical mass many insurance companies have merged. Indeed, the larger an insurance company, the more risks it can accept, which in turn could be an important factor influencing performance. Insurance companies also have a substantial proportion of fixed costs, which creates opportunities for exploiting economies of scale. Therefore *size* is an important variable for both the insurance and the banking industry.

It seems that the Dutch groups are in a better position when considering this factor. Apart from the mergers between banks and insurance companies, Dutch banks have also engaged in mergers and acquisitions. The most notable example is the merger between ABN Bank and AMRO Bank. These banks were both among the largest in the Netherlands, and when they merged they not only became the largest bank in the Netherlands, but also the largest foreign bank in the U.S.

The largest Belgian banks are generally smaller than their Dutch counterparts. However, last year a discussion of the *Large Belgian Bank* emerged. There have been talks between the managers of three large banks about a potential merger, but in the meantime the idea has been blown the whistle due to diverging viewpoints of the main shareholders and the integration of one of the largest banks, BBL, into the ING Group. This also applies to the insurance industry: Dutch insurers are generally larger than Belgian insurers.

Some financial institutions have tried to find expansion in foreign markets. Although the new merger wave towards financial conglomerates has given a new impulse to this issue, *geographical diversification* has been a fact for decades. For a long time, both banks and insurance companies have had foreign subsidiaries. At first these subsidiaries were situated in the neighbouring countries; but later on expansion towards other countries was pursued. Again, it seems that the larger Dutch financial firms have been more active:

ING, Aegon, ABN-AMRO, Rabobank and Fortis (which is partially a Belgian conglomerate) have considerable businesses outside their home country. As already mentioned, some of these new financial conglomerates have actively pursued an internationalisation strategy and have combined it with their diversification strategy. This is the case for the Dutch ING Group who considers the development of the all finance strategy in greenfields as an element of their core strategy.

A last important variable is the *corporate governance structure* of financial institutions. In our sample there are large governance differences. In some smaller companies the functions of owners and managers are integrated; in other companies the gap between owners and managers is much larger, e.g. because of the creation of *structure companies* (see Kabir e.a., 1997). A large part of the financial institutions in our sample are subsidiaries of foreign financial firms. Sometimes these subsidiaries are restricted in their strategic actions (e.g. they have no opportunities for geographical expansion if the group is present in other countries as well). On the other hand, being a subsidiary of a large international group may also be advantageous (e.g. in attracting the necessary funds at a better rating -e.g. the upgrading of the rating for BBL after the integration into the ING Group).

As a conclusion, we can state that the Dutch and Belgian financial institutions differ on a series of characteristics:

- differences in product diversification, not only in terms of relative importance (content) and mode of diversification, but also in the way diversification is implemented (process);
- differences in size, in terms of total assets and turnover;
- differences in the degree of internationalisation (geographical diversification);
- differences in corporate governance structure.

The purpose of this paper is to test to what extent these factors determine performance. Thereby, we will primarily focus on the effects of diversification because financial conglomeration is our main topic of interest. This discussion will take into account the peculiarities of the financial services industry.

HYPOTHESES ON THE DIVERSIFICATION EFFECTS

We have spent considerable attention to the concept of synergies (see paragraph 3.3). As already mentioned a frequently used operationalisation is looking at the strategic relatedness. If synergies between banks and insurance companies may constitute a competitive advantage of financial firms, then financial conglomerates have an advantage compared to specialised banks and insurers:

Hypothesis 1: Financial conglomerates outperform 'pure' banks and 'pure' insurance companies.

Researchers have found a relation between acquisition success and size of the target company. Consistent with the idea that the relative size of the compounding parts determines acquisition performance, we argue that the relative share of banking and insurance subsidiaries also determines the amount of synergy that can be exploited. This in turn is reflected in the performance figures of the financial institutions:

Hypothesis 2: Financial conglomerates with relatively equal insurance and banking parts profit more from the benefits of synergies. When there is asymmetry, the gains are smaller.

The main contribution of this chapter is that it analyses whether different forms of integration have an effect on a firm's performance. We have argued that organisational fit is as important as strategic fit. This is especially true when mergers and acquisitions (but other entry vehicles as well) are set up to benefit from synergies between the compounding parts. We have made a distinction between management synergies, marketing synergies, production synergies, human resource management synergies and financial synergies. Furthermore we propose that the mode of diversification also contributes significantly to diversification success.

We refer to the paragraph 3.3.3 and 3.3.4 for the theoretical arguments as why we included these issues into our analysis. The arguments formulated above are tested by the following hypotheses:

Hypothesis 3a: Different types of synergies will have a different effect on banking firm performance. Furthermore the mode in which banks have entered the insurance sector is an important determinant of bank performance.

Hypothesis 3b: Different types of synergies will have a different effect on insurance companies' performance. Here again, the mode of entry is of decisive influence on the performance.

A firm's performance is not only influenced by the degree of diversification. There are other control variables as well.

One of the most important control variables is the *industry* in which a firm operates. Although we have opted to study the financial services industry, a distinction can be made between the banking and the insurance sector. Within the insurance industry the life market differs quite substantially from the non-life business. Thus it is still relevant to include industry factors in our analysis. The discussion on the importance of industry variables traces back to the first studies on diversification: the better performance figures of related diversifiers in Rumelt's study were also attributed to the inclusion of firms of highly profitable industries in his related category. From then on, a lot of studies have addressed the question: "what drives profitability?" (Rumelt, 1991; Wernerfelt and Montgomery, 1988; Roquebert e.a.; 1996). Nowadays, it is assumed that both industry and organisation factors have a valuable contribution in explaining firm performance:

Hypothesis 4a: Firm performance is positively related to industry profitability.

As we have shown in the section on the European financial sector (point 6.2), *size* is another important controlling variable. Both the banking and the insurance industry are subject to horizontal mergers because an increase in scale is seen as a necessary condition for survival. Johnson and Thomas (1987) have argued that larger firms also have more stable income streams. In this way, size serves as a proxy for risk. Nguyen e.a. (1990) used firm size to account for efficiency differentials (scale economies) which in turn are considered as entry barriers for a particular industry:

Hypothesis 4b: Firm performance is positively related to size.

Internationalisation which can be considered as a kind of geographical diversification can be pursued if a firm competes in maturing markets. Diversifying into other markets may be seen as one of the easiest ways to guarantee further growth. Several researchers have revealed a positive relationship between internationalisation and firm performance (for a review see Sambharya, 1995). Internationalisation also contributed to stable corporate profits. Some authors (Kim, Hwang and Burgers (1989); Geringer e.a. (1989); Sambharya (1995)) have also investigated the combined effect of product and market diversification and found significant results for the interactive effect:

Hypothesis 4c: Firm performance is positively related to internationalisation.

An important moderating variable in the European financial services industry is the *governance structure*. In Europe 'corporate governance' has become one of the buzz words of the moment and has received a lot of attention both from the academic as well as from the business world. American research has shown that the governance structure is an important variable for several reasons: Hill and Snell (1988) showed that in research-intensive industries stockholder concentration had a significant impact on strategy (diversification vs. innovation). If stockholders were relatively concentrated, an emphasis upon innovation was favoured which in turn was associated with higher profitability. This is consistent with agency theory which predicts that managers rather pursue growth than profitability and thus have other goals than stockholders. A similar research project was performed by Kroll e.a. (1997) who investigated whether acquisitions were driven by the interests of shareholders or CEOs. The results indicated that acquisitions of *owner-controlled* and *owner-manager-controlled* firms (i.e. there are significant shareholders who closely monitor the strategies of top managers) were evaluated positively by the market:

Hypothesis 4d: The more stockholders are able to direct a firm's strategy, the higher the profitability.

Lastly, we also want to investigate the impact of *risk* on performance. Conventional wisdom suggests that increased risk will be reflected in increased performance. However this is not always the case. Miller and Bromiley (1990) indicated some studies with opposite findings of which Bowman's were the most interesting. Bowman introduced the *prospect theory* which starts out from the idea that firms compare their performance to the average performance in the industry. When the firms notice that they underperform, they are prepared to take more risks in order to increase their performance figures. A firm outperforming its industry peers is rather risk-averse and is only prepared to take risks if an investment opportunity offers high expected returns.

Several researchers have also investigated the link between diversification, risk and performance. Early research discovered a negative risk-return association for related diversifiers, while for unrelated diversification a positive risk-return relation was found. Chang and Thomas (1989) have taken into account other moderating variables as well and found that the influence of the diversification strategy might change the risk profile of a firm, but the effects were not so important when compared to size effects, diversity effects and market effects. They also found confirmation of Bowman's prospect theory:

Hypothesis 4e: Firm performance is positively associated with risk.

METHODOLOGY

Definition of financial conglomerates

Empirical research on financial conglomerates is very scarce for several reasons: (1) financial conglomerates are a relatively new phenomenon; (2) there is still no uniform regulatory framework for these financial conglomerates. Furthermore, there are still different definitions of financial conglomerates which are sometimes conflicting. Therefore, we have opted to choose for a broad definition of financial conglomerates by defining them in the way described in chapter 1:

> "A financial conglomerate relates to a group of enterprises, consisting of different kinds of financial institutions..." (Van den Berghe, 1995: 2).

What is essential is that there are at least some capital links between a bank and an insurance company (whether via a joint venture, an acquisition or via the creation of a subsidiary). When a bank had an insurance broker at its disposal, we did not consider this group as a financial conglomerate because banks have acted as broker for decades. Furthermore it is very difficult to extract the figures for these subsidiaries.

Sample

Since no publicly available databases exist, it is rather difficult to compose a database on financial conglomerates. However, the Erasmus Finance and Insurance Centre (EFIC), a department of the Erasmus University of Rotterdam involved in strategic research on the financial services industry, has done a good job by setting up a private database which was made possible by the Levob Foundation and which identifies the capital links between the main (European) financial institutions. We have used this database to detect the largest Belgian and Dutch financial conglomerates. Furthermore, we have also included the largest non-diversifying banks and insurance companies which acted as control group for the financial conglomerates. So our unit of analysis is the Belgian or Dutch mother company of a (group of) financial institution(s). This does not always mean that this is the ultimate mother company of the group. This is the case for the subsidiaries of large international groups with their head office outside Belgium or the Netherlands. Then we look at consolidated figures for the domestic organisations. For example, Delta Lloyd is the Dutch insurance holding of the British General Accident-Commercial Union Group. Sometimes we have analysed the Belgian and the Dutch subsidiary of an international group. For example, this is the case for AXA Belgium and AXA Leven -both are subsidiaries of AXA, the largest European insurance holding- but because there are no operational (and direct capital) links between the two subsidiaries and financial data is available, we treat them separately. Furthermore we restricted our sample to Belgian and Dutch financial firms because we wanted to analyse non-standardised data on diversification, which was not available in other databases. Some large financial groups were removed from the sample due to a lack of data, but generally the largest groups are included. Table 6.1 provides an overview of our sample.

We composed our sample taking into account the 1995 situation. However, mergers and acquisitions have been going on, which explains why some of these groups do not exist any more, at least in the form in which we have analysed them. This is the case for Group Josi, which has become a subsidiary of Winterthur Europe; furthermore, Mercator Verzekeringen and Noordstar Verzekeringen were involved in a merger, etc.

Table 6.1 Overview of the sample

Financial conglomerates (21)	
ABN AMRO Holding (NL)	Gemeentekrediet-OMOB (B)
Achmea (NL)	Generale Bank (B)
AEGON (NL)	ING Group (NL)
Almanij-KB Group (B)	Ass. Groep Josi-BVB (B)
BACOB-DVV (B)	Ohra Bank & Verzekeringsgroep (NL)
BBL (B)	P&V-Bank Nagelmackers (B)
CERA-ABB (B)	Rabobank-Interpolis (NL)
De Vaderlandsche (B)	Reaal Groep (NL)
Delta Lloyd (NL)	Royale Belge (B)
Fortis AG (B)	SNS-NOG (NL)
Fortis AMEV (NL)	
Banks (15)	
Anhyp (B)	Deutsche Bank de Bary (NL)
Bank of Tokyo-Mitsubishi (NL)	Europabank (B)
Banque Paribas (NL)	Friesland Bank (NL)
BEAL (B)	Indosuez Bank België (B)
Central Hispano Benelux (B)	NIB (NL)
CLBN (NL)	Paribas Bank Belgium (B)
Crédit Lyonnais Belgium (B)	van Lanschot Bankiers (NL)
Dai-Ichi Kangyo Bank (NL)	
Insurance companies (13)	
AXA Belgium (B)	UAP-Nieuw Rotterdam (NL)
AXA Leven (NL)	Univé Verzekeringen (NL)
Generali Belgium (B)	Winterthur Europe (B)
Goudse Verzekeringen (NL)	Zürich Nederland (NL)
Mercator Verzekeringen (B)	Zwitserleven (NL)
Noordstar Verzekeringen (B)	Zwolsche Algemeene (NL)
Royal Nederland (NL)	

(B): Belgian - (NL): Dutch

This research project focuses on the impact of diversification strategies on corporate performance. More and more, diversification researchers (Davis and Thomas, 1993; Hoskisson and Hitt, 1990; Brush, 1996) argue that the appropriate unit of analysis is the business unit because synergies provide business units with a competitive advantage. We do not deny this argument but we state that if business units benefit from synergies and integration, we think that this should also be reflected in the corporate performance.

Diversification variables

Consistent with the recommendations made by methodologists who have recognised the importance of using multiple measures of theoretical constructs (Davis e.a., 1992), we have used different measures to approach diversification. Datta e.a. (1991) have distinguished between the degree, type and mode of diversification. We use degree, mode and process of diversification. We neglected the type of diversification (related vs. non-related) because our main purpose was to investigate whether financial groups who actively exploit the synergies between their banking and their insurance subsidiaries outperform the non-diversifying financial institutions. Furthermore, financial institutions were subject to regulation which has prevented diversification for a long time.

In order to obtain the *degree of diversification* (DIV), we calculated the proportion of total assets and the proportion of total income generated by the new business for each year within the 1991-95 period. Generally banks have larger total assets, insurance companies have a larger total income because gross premium income is included. We then averaged these two numbers because in this way we filtered out the effects of being a bank or an insurance company. The more equal the relative size of the two compounding parts, the larger the diversification measure. It is easy to deduct that the range of this variable is [0; 0.5].

When operationalising the *mode of diversification* (MODE - see appendix 1), normally a distinction is made between internal development and mergers and acquisitions. We have argued that besides these two categories joint ventures should also be included. We have added another category which is quite typical for the financial services industry. Sometimes, banks and insurance companies have been part of the same group for a long time without pursuing synergies. Now that the major competitors have engaged in one or another form of financial conglomeration, these firms are looking inside the group to set up co-operation agreements.

The operationalisation of the *process aspects* of diversification has taken a long time. We had to develop a framework for the financial services industry which differs -like any other service industry- remarkably from the manufacturing industry. Because of the peculiarities of (financial) service firms general frameworks which have mainly originated from the manufacturing industries, are not always directly applicable to these firms. Based on Porter's value chain and the servuction system of Langeard, we constructed the *'financial conglomerates control board'* (see chapter 5). This reference scheme was presented to academicians and managers of financial conglomerates[7] for evaluation. These discussions provided us with useful insights as to which activities were the most important 'drivers of synergy'. According to Porter, every value activity can be the base for the exploitation of synergies (1985: 326), but the discussions we had, revealed that managers of financial conglomerates focused on particular sets of synergies.

---

7 We first presented this F.C.C.B. for an audience of experts at a conference in Brussels in March 19th & 20th of 1996. Affterwards we conducted experts interviews with several managers from the following financial groups: ING Group, Achmea, Almanij-KB Group, SNS-NOG Group and Fortis (AMEV).

We made a distinction between five categories of synergies: (1) financial synergies (CAPINT), (2) management synergies (MANINT), (3) human resource synergies (HRMINT), (4) synergies on production (PRODINT) and (5) marketing synergies (MKTINT). The main point was to identify to which degree a financial conglomerate strives for a particular type of synergy. There we introduced the concept of *integration*: we theorised that if a conglomerate integrates e.g. the top management, this is a sign that they pursue management synergies. We had to develop different classification schemes for the different types of synergies (see appendix 1) since no standard instruments were available. The financial newspapers and magazines were the main data source used to compile this database. Afterwards we sent out a small questionnaire to executives in order to complete our database and to check for remaining questions. In this way we assembled data on all integration measures for each year within the 1991-1995 period.

Performance measures

A quantitative analysis of performance effects of diversification strategies also requires a comprehensive set of financial data of the financial conglomerates and the firms in the control group. It is generally accepted that strategic research should take into account different measures of performance. However, we relied solely on accounting data mainly because market-based data were not available for all cases. We acknowledge that there are deficiencies in the accounting data but by deflating our figures we hope to find a solution for the economic distortions. Managerial manipulation is more difficult to resolve and there we adopted the view that the correlation between 'real' performance and reported performance is high, i.e. if firms are performing well this should be reflected in the annual accounts. We examined the 1991-95 period: as a result we disposed of 5 observations for each case.

A lot of mergers occurred in the beginning of the nineties, so we expected that the effects of these strategic decisions were reflected in these figures. It could be that the effects are not fully incorporated in these figures, because in mergers and acquisitions expenses must be incurred before profits can be gained.

Contrary to the case of industrial firms, there are no official databases with financial data for financial institutions. Therefore we had to rely on annual accounts and privately maintained databases. In this respect, we would like to thank Bureau van Dijk who offered us their most recent version of *Bankscope*, a CD-ROM with the statements of 9,200 world banks. For the insurance companies we made use of publications of Coopers and Lybrand (*Trends bij Verzekeringsmaatschappijen*) for Dutch insurance companies and the *Financiële Gids van de verzekeringen*, a Kluwer Editorial publication of statements of Belgian insurance companies. For all cases we disposed of consolidated data, except for the Belgian insurance companies. There we consolidated all Belgian subsidiaries. However data of foreign subsidiaries was not always available. For example we were not always able to include the Luxembourg subsidiaries who experienced a remarkable growth in 1995.

We have consolidated performance figures for the banking and the insurance subsidiaries of the financial conglomerates and their industry peers. This allows us to investigate the impact of diversification both in the banking and the insurance industry. We calculated average ROA and ROE before tax because profit before tax better reflects operating income. Furthermore, we calculated the compound sales growth (CSG), which serves as a measure of the growth of total income, and an efficiency measure (EFF), which we defined as the ratio of operating costs to total income.

Control variables

In order to control for the *industry* in which a firm is active, we developed two industry-weighted return variables (based on either ROA before tax -INDROA- and ROE before tax - INDROE). As already mentioned we divided the financial services industry in three sub-sectors: (1) the banking industry, (2) the life insurance industry and (3) the non-life insurance industry. For each industry we disposed of the average ROAbt and ROEbt and dependent on the proportion of each activity, we calculated a firm's industry-weighted return for each year within the 1991-95 period.

Size (SIZE) has been operationalised by calculating the average of two measures: (1) natural logarithms of average total assets and (2) natural logarithms of average of total income. We computed this average to eliminate the effects of being a bank or an insurance company: generally banks have larger balance sheets while insurance companies have larger total income.

For *internationalisation* (INT), we computed the average proportion of domestic activities (a measure based on assets, total income and number of employees) on total activities. Then internationalisation should rather be interpreted as the complement of our measure, thus a low measure indicates high internationalisation, and a high value for INT is an indication for high domestic activities.

Normally, *corporate governance* is measured by looking at the stake of the largest shareholder. We have developed a classification scheme (CGCAT - see appendix 2) which better reflects the current situation in the European financial services industry. Again, the main dimension is the degree to which managers (as opposed to stockholders) have control over the (corporate) strategy.

Lastly, the *risk* measure was computed by calculating the ratio capital to assets. It was hypothesized that the better capitalised the financial firm, the less risky it was (thus, a high ratio indicates a firm with relatively low risk). In this way it was possible to compute a risk measure for each year separately. This is not the case if one uses a risk measure based on standard deviations and variability of returns.

Table 6.2 List of the variables

Performance variables			
ROAbt	Income before tax/total assets	ROEbt	Income before tax/capital
CSG	Compound sales growth	EFF	Operating efficiency
Diversification variables			
DIV	Average degree of diversification	MKTINT	Integration of marketing activities
CAPINT	Integration of capital	HRMINT	Integration of HRM
PRODINT	Production integration (i.e. integration of the investment activities)	MANINT	Integration of top management teams
Control variables			
SIZE	Average of natural logarithm of total assets and total income (1991-95)	CGCAT	Corporate governance measure defined as degree of control exercised by managers
S.D. ROAbt	Corporate risk defined as standard deviation of ROAbt over 1991-95	S.D. ROEbt	Corporate risk defined as standard deviation of ROEbt over 1991-95
INT	Proportion of domestic activities on total activities	RISK	Capital/total assets
INDROAbt	Industry-weighted ROAbt	INDROEbt	Industry-weighted ROEbt

PRELIMINARY RESULTS

1. In table 6.3 we have compared some performance measures for the different categories of financial institutions. We used an analysis of variance (ANOVA) to detect significant differences between the performance measures of these sub-groups.

When looking at the general performance (average ROAbt and ROEbt), the results of the analysis of variance are somewhat mixed. If we take ROEbt as measure of performance, no significant differences between financial conglomerates, banks and insurance companies are found. However, when considering ROAbt then insurers outperform their banking and conglomerate peers. However, we should take into account that this performance measure is somewhat biased because generally banks and financial conglomerates have larger balance sheets. These results suggest that the leverage effect is an important explanatory factor for the differences found[8].

[8] Leverage is a measure of financial risk and can be defined as the ratio of ROEbt by ROAbt. If we compute this leverage effect for the different financial

Table 6.3 Analysis of variance for the different financial institutions (n = 49)

Variables	Financial congl.	Banks	Insurance comp.	F-value
ROAbt	.0070	.0053	.0120	4.293**
ROEbt	.1150	.1174	.1223	.025
CSG	.0890	.0360	.0860	3.447**
EFF	.9062	.7634	.9683	21.276***
SIZE	8.637	6.114	6.405	30.785***
S.D. ROAbt	.0020	.0015	.0060	6.457***
S.D. ROEbt	.0345	.0283	.0467	1.252
CAP/ASSETS	.062	.041	.127	12.399***

*, **, ***: *statistically significant resp. at the .1, .05 and .01 level*

For both ROAbt and ROEbt, non-diversifying insurance companies have the best performance figures. In other words hypothesis 1 is not confirmed by our data.

We looked at other performance measures as well. For what concerns sales growth (CSG), banks score significantly less than financial conglomerates and insurance companies. On the other hand banks have a relatively good efficiency (EFF). Due to the growing competition banks have put pressure on efficiency for a long time. The competition in the insurance industry is of a more recent nature which explains partly why their efficiency is much lower. Furthermore there are still some insurance businesses in which the technical results are very bad which in turn increases the operating costs.

The generally accepted hypothesis that financial conglomerates are less risky than pure banks or insurance companies is not confirmed. Whether we use standard deviations of returns (S.D. ROAbt) or the proportion of capital on total assets (CAP/ASSETS), financial conglomerates do not outperform pure banks and insurance companies. However, it is noteworthy that different indicators of risk result in different findings. When considering risk as the variation in income stream, banks have the best risk profile. The variation in their returns is the smallest (both for ROAbt as for ROEbt, but only in the first case the results are significant). However when we look at risk from a solvency point of view, we see that insurers have much more capital relative to total assets. We have already indicated that the leverage effect which is assumed to be a measure of financial risk should also be taken into account.

2. In a second phase we examined whether the *relative size* of the compounding parts could shed a light on the relation

institutions, we find major differences: financial conglomerates: 16.42; banks: 22.15 and insurance companies: 10.19.

diversification-performance. We hypothesised that more equal banking and insurance parts allowed for more synergies. This in turn should be reflected in superior performance. Therefore, we ran regressions on both ROAbt and ROEbt, testing for both a linear and a quadratic function. If we include all observations (5 year observations for 49 cases), we obtain the results described in table 6.4.

Table 6.4 Regression of measures of performance on diversification (n = 245)

	ROAbt	ROEbt	ROAbt	ROEbt
Intercept	.0083***	.1223***	.0083***	.1223***
DIV	-.0083	-.0786	-.0011	-.0838
DIV^2			.0305	.0220
F-statistic	2.294	1.291	1.199	.643
R^2	.0093	.0053	.0099	.0053
Adj. R^2	.0053	.0012	.0016	-.0029

*, **, ***: *statistically significant resp. at the .1, .05 and .01 level*

No significant relationships were found and the proportion of the explained variance is almost equal to zero. When we restrict the sample to the financial conglomerates (n = 100), we obtain the results in table 6.5.

Table 6.5 Regression of measures of performance on diversification (n = 100)

	ROAbt (1)	ROEbt (2)	ROAbt (3)	ROEbt (4)
Intercept	.0080***	.1385***	.0072***	.1508***
DIV	-.0067	-.1574**	.0134	-.4729*
DIV^2			-.0716	1.1248
F-statistic	2.232	5.411	1.821	3.4769**
R^2	.0223	.0523	.0362	.0669
Adj. R^2	.0123	.0427	.0163	.0476

*, **, ***: *statistically significant resp. at the .1, .05 and .01 level*

The regressions (2) and (4) (both on ROEbt) yield significant coefficients for the explanatory variable. We find weak evidence for the statement that diversification reduces value. This statement is contradictory to hypothesis 2, where we expected that higher diversification would result in higher overall performance.

3. In the preceding chapters we stressed the importance of organisational fit for acquisition success. We have operationalized organisational fit by looking at measures of integration for different activities.

We hypothesised that the reasons why banks engage in insurance activities may be different than why insurers offer banking products. Therefore we think that it may be useful to consider integration in both the banking and the insurance industry separately. We ran several regressions of which the results are presented in table 6.6.

Table 6.6 Regression of measures of performance on types of synergy

	Banking industry (n = 100)		Insurance industry (n= 100)	
	ROAbt	**ROEbt**	**ROAbt**	**ROEbt**
Intercept	-.0007	.0773**	.0067	.0678
MODE	-.0012	.0239*	.0020	.0134
CAPINT	.0007	-.0302***	-.0014	-.0205
MANINT	-.0004	.0081	.0012	.0197
HRMINT	-.0003	-.0013	-.0010	-.0124
PRODINT	-.0003	-.0056	.0027***	-.0016
MKTINT	.0030***	.0330***	-.0009	.0270
F-statistic	2.578**	3.875***	4.103***	1.772
R^2	.1426	.2000	.2093	.1026
Adj. R^2	.0873	.1484	.1583	.0447

*, **, ***: *statistically significant resp. at the .1, .05 and .01 level*

As in table 6.5, we restricted the sample to the financial conglomerates. Non-diversifying banks and insurance companies have a zero-score on the integration items and were left out of the analysis because of potential bias of our results.

In general, we do not find evidence for the hypotheses we proposed. If we look at the regressions for the *banking subsidiaries* of the financial conglomerates, we see that integration with the insurance departments of the group does not result in significant improvements of both return on assets and return on equity. The regressions reveal that only integration on the marketing level has a positive impact on performance but is counterbalanced by capital integration (at least in the second regression). On the other hand, the regression on ROEbt reveals that the way in which banks have entered the insurance industry (MODE) is an important determining factor. Banks do better by setting up their own subsidiary, when they engage in looser forms of co-operation.

For the insurance subsidiaries, the results were more difficult to interpret and no conclusive statements could be made. Not only the variable 'mode' but almost all integration variables are insignificant. The only exception is the production integration variable for the ROAbt regression: there are advantages for insurance companies to integrate the investment activities with those of their banking partners.

The problem in the regressions of table 6.6 is multicollinearity: the integrated financial conglomerates try to achieve synergies on all types of synergies, thus all integration categories are highly correlated. We have tried to resolve this problem by capturing MODE, CAPINT, MANINT and HRMINT into one measure, indicating the structural integration but there was no clear improvement in the results.

Overall we can conclude that organisational fit gives no additional insight into the relation between diversification and performance.

4. Ultimately, we wanted to check the influence of some moderating variables. Therefore, we regressed the two performance variables ROAbt and ROEbt on the diversification variable and control variables for size, industry, corporate governance structure, internationalisation and risk. First, we give a descriptive analysis of the main variables (table 6.7), then the results of the regression are presented in table 6.8.

The most remarkable difference between the two regressions in table 6.8 is their explanatory power. The regression on ROAbt has a much larger adjusted R-squared and thus gives a better explanation for the variances in performance. However, there are some interesting similarities in the coefficients of the independent variables.

Generally, diversification and size do not explain much of the differences in performance. For both regressions diversification has a negative coefficient but remains insignificant. The impact of size on performance is negligible. As we have indicated, the pace of mergers and acquisitions has increased substantially in the financial services industry. The primary objective for these takeovers is to benefit from an increase in scale and scope. The results we obtained suggest that these strategic decisions might not always lead to the desired effects.

No significant relationship was found between the degree of internationalisation and performance. However when we did a two-variable regression (performance on internationalisation) we found a significant positive relation between the degree of internationalisation and performance. The reason why internationalisation has become insignificant is perhaps because of the presence of multicollinearity (see table 6.7).

Table 6.7 Descriptive statistics of the main variables

	Mean	S.D.	Correlations							
			1	2	3	4	5	6	7	8
1. Performance (ROAbt)	.008	.007								
2. Performance (ROEbt)	.117	.101	.467**							
3. Size (SIZE)	7.846	1.996	-.281**	.135*						
4. Diversification (DIV)	.059	.093	-.097	-.073	.456**					
5. Internationalization (INT)	.887	.222	-.250**	-.108	-.074	-.213**				
6. Industry-weighted performance (INDROAbt)	.011	.005	.299**	-.016	-.197**	.217**	-.148*			
7. Industry-weighted performance (INDROEbt)	.142	..036	-.298**	.232**	.346**	-.179**	.185**	-.330		
8. Management control (CGCAT)	2.347	1.19	.134*	.246**	.373**	-.002	-.528**	-.114	.098	
9. Risk (CAP/ASSETS)	.074	.059	.562**	-.154*	-.635**	-.133*	-.319**	.407**	-.653**	-.136

*, **: *Pearson correlation is significant at the .05 resp. .01 level (2-tailed)*

Table 6.8 Regression of performance measures on diversification and control variables (n = 245)

	ROEbt	ROAbt
Intercept	.0363	-.0091
Diversification	-.0581	-.0043
Size	-.0001	.0003
Internationalisation	-.0302	.0033
Industry-weighted return	.5529*	.1820*
Management control	.0159**	.0017***
Risk	-.0469	.0832***
F-statistic	4.775***	23.984***
R^2	.1074	.3768
Adj. R^2	.0850	.3611

*, **, ***: *statistically significant resp. at the .1, .05 and .01 level*
All regressions were corrected for heteroskedasticity - White heteroskedasticity-consistent standard errors and covariance

There are two variables that have a significant impact for both the ROEbt regression and the ROAbt regression. First, the positive coefficient for the industry-weighted return indicates that the industry in which a firm operates has a decisive influence on a firm's performance. If the industry is doing well, then this is reflected in a firm's performance. These results are in line with findings of Wernerfelt and Montgomery (1988) and Rumelt (1991).

Another important variable in the financial services industry is the corporate governance variable *management control*. Our findings indicate that the performance of a financial institution is positively related to the degree to which managers have control on the corporate strategy. It is generally accepted that the more the stockholder is able to control the managers and the more aligned the goals of the manager and the stockholder, the better the performance. Our findings point out the opposite: when a manager is able to develop the corporate strategy a firm may benefit more than when stockholders have the power of determining the strategy. This could be an indication that managers of financial institutions may pursue rather long-term strategies while stockholders are more interested in short-term profitability thereby neglecting strategic opportunities.

In the regression on ROAbt we also found evidence for hypothesis 4e: there is a positive significant relation between risk and return on assets. This is consistent with the idea that an investor only wants to take more risk if a higher return is achieved.

DISCUSSION

The main contribution of this chapter is that it has provided some empirical evidence on a new phenomenon in the financial services industry, namely the emergence of financial conglomerates. We have tried to develop an industry-specific framework, taking into account the recommendations made by other researchers in this area. In this way, this research is also interesting from a theoretical point of view because it investigates the effects of different elements of the diversification strategy on corporate performance, as suggested by prominent diversification researchers.

However, the emerging picture is not very clear. The theoretical advantages ascribed to diversified financial groups are not (directly) reflected in their performance figures. Table 6.3 shows that there are significant differences between financial conglomerates, pure banks and pure insurance companies. The non-diversifying insurance companies are most profitable. On the basis of ROEbt financial conglomerates score the least, if ROAbt is our performance measure, then financial conglomerates have a better performance than banks. The figures are in clear contrast with the hypothesis we proposed that due to synergies between banks and insurance companies, financial conglomerates should outperform non-diversifying financial institutions. As we already mentioned, it could be that the effects of diversification are not fully incorporated in our performance figures because it takes time and money to implement a diversification strategy successfully. On the other hand, many conglomerates originated in the beginning of the nineties. In other words we expect that some effects of diversification are at least incorporated in the 1994 and 1995 figures.

The differences between financial conglomerates, banks and insurance companies are not only restricted to ROAbt and ROEbt, but are also reflected in average size, compound sales growth, efficiency and risk.

When we investigated the relation diversification-performance somewhat further, we found a (weak) negative relationship between the degree of diversification and the ROEbt. Our findings do not fully confirm the propositions made by Jones and Hill (1988) who analysed diversification from a transaction cost point of view. Jones and Hill made the distinction between related and unrelated diversification and came to the following profit functions for related *(Pr)* and unrelated *(Pu)* diversification. We have added the profit function of the financial conglomerates *(Pfc)* to the picture.

Jones and Hill argue that there is an optimal level of diversification: when the level of diversification is small, there is room for further expansion. But once a certain point is reached, then the costs of diversification outweigh the benefits and further diversification becomes less profitable. Our findings suggest that the optimal level of diversification is *no diversification*. We have tested for a quadratic function but found no proof in our data set.

Figure 6.1 Profit functions derived from Jones and Hill (1988)

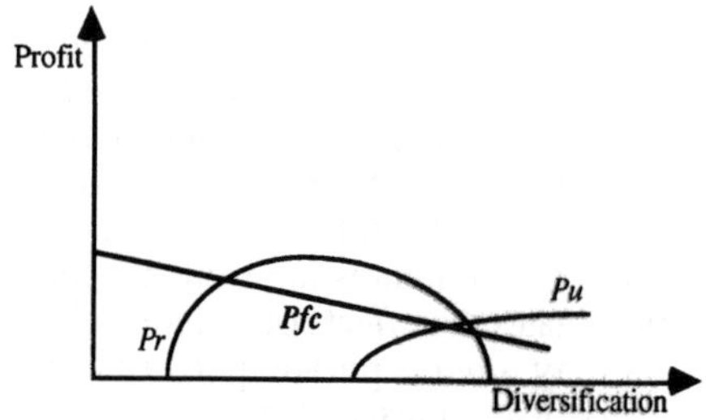

At least this applies to the *group* of financial conglomerates studied. One explanation for the difference between successful diversification at the micro level and negative performance effects on the sub-industry level could well be that diversification is a difficult strategy that leads to more losers than winners.

A central topic in this paper is the process of diversification. We focused on the way a financial conglomerate integrated different value activities. Literature about this subject is still in an emerging phase and most researchers specify only one or two process variables. Like Mahajan and Wind (1988) we have taken a broader view and we included more categories: capital synergies, management synergies, synergies on human resources management, production synergies (in the form of integration of the investment department) and marketing synergies. Analogue to what Mahajan and Wind (1988) found, we also discovered that not all types of synergy result in positive gains.

Furthermore, the effect of the different types of synergy was different according to the industry in which the firm operated. For example, in the banking industry capital integration resulted in negative performance. However this negative effect was offset by the positive effects of marketing integration. For the insurance industry, integration of the investment activities yielded positive performance figures. Thus the advantages of a combination of long-term oriented savings of insurance companies and the short-term oriented savings of banks are rather reflected in the profit figures of the insurance companies.

Overall there is no clear pattern of which activities should be integrated: what is good for the banking businesses is not always good for the insurance companies and vice versa. In practice a distinction can be made between highly-integrated financial conglomerates and conglomerates with a low integration. Based on the outcome of all four regressions the performance of highly-integrated financial conglomerates is higher than when the banking and insurance subsidiaries strive for autonomy and no co-operation. At least this is an indication that organisational fit plays a (perhaps modest) role in the diversification debate.

A last regression revealed the impact of diversification on performance compared to other controlling variables. As such, there are more important variables than diversification in explaining the variations in performance. Conform Schmalensee, Wernerfelt and Montgomery (1988) and Roquebert e.a. (1996), we also found that the industry in which a firm operates has major influence on a firm's performance. Also interesting is that the governance structure plays an important role in explaining performance differences: performance gets better when managers have more control over a firm's corporate strategy. This statement is contradictory to the generally accepted hypothesis that good performance is associated with stockholders' ability to control for a firm's strategy. One reason for these findings is that managers may pursue long-term strategies, while stockholders focus on the short-term profitability. Our results suggest that a focus on the long-term may determine performance in a positive way.

We conclude this chapter by indicating some of the *major shortcomings* of this research project and with some recommendations for future research. We have focused on *diversification* between banking and insurance activities and we have developed several measures for this diversification strategy, highlighting on different aspects of this strategy. However banking activities should not be considered as one homo-geneous set of activities and thus, banks can achieve diversification within their industry. For example, a lot of banks try to increase their non-interest income (fees) in order to be more independent from interest rate movements. Therefore more and more banks engage in investment banking, corporate banking, etc. as opposed to their retailing activities. In this chapter we were not able to distinguish between these different activities because no financial data was available. The same holds for the insurance industry: generally distinction is made between life and non-life (general) business. But within these categories, there are different subcategories. We only considered the distinction between life and non-life insurance to determine industry-weighted returns.

A second element which requires further examination is the *unit of analysis*. While we have focused on the corporate level, arguing that diversification is a matter of corporate strategy (and should therefore be evaluated on this level), we acknowledge the importance of synergies on the business unit level. However, other types of synergy and other measures of synergy may be more appropriate if the unit of analysis becomes the business unit. Furthermore, the stress on more intangible kinds of resources and the importance of capabilities is only marginally included in our analysis. On the other hand, we have used some predefined and objective items for defining the degree to which synergies were pursued which is an important factor when considering the validity.

A third drawback of this study is the *use of accounting-based performance measures*. The reason why we used them is quite straightforward (see methodology section on performance variables), but the inclusion of other performance variables may be a valuable addition. First, we think of including market-based measures, but equally important are more operational performance measures. For example when linking different cost measures and different types of synergies we could establish more direct relationships between type of synergy and performance.

APPENDICES

Appendix 1: Operationalisation of the synergy variables

A. *Mode of diversification*

1.	Start-up / internal diversification	4
2.	Mergers and acquisitions (> 1990)	3
3.	Financial firms belong to the same group but have recently discovered co-operation possibilities	2
4.	Co-operation agreement between two independent partners → the co-operation is formalised (e.g. in the form of a joint venture)	1

B. *Capital structure → financial synergies*

1.	Mainly 100% participations between the main subsidiaries of the group	5
2.	Mainly majority participations (on the average between 50% and 90%)	4
3.	Important portion of non-majority holdings but nevertheless 'control'	3
4.	Mainly 'substantial stake(s)' (20% - 50%) but without direct control	2
5.	Participation(s) between main companies of more than 5%	1

C. *Mangement integration → management synergies*

1.	There is one common, unified top management team	5
2.	The management team of the subsidiary is controlled by the management team of the mother company	4
3.	Reciprocal representation in each other's management team	3
4.	Reciprocal representation in each other's board	2
5.	Unilateral representation in the board	1

D. *Integration of human resources → flexibility*
Each condition which is met, adds one point to the integration score.

1.	Co-operation between the different HRM departments	1
2.	CAO per industry	1
3.	Centrally co-ordinated personnel policy	1
4.	Central Work Council	1
5.	Job rotation	1

E. *Production integration → integration of the investment policy*

1.	Own organisation to co-ordinate and centralize the financial policy	5
2.	Co-operation in a structured way	3
3.	Ad hoc co-operation	1
4.	Different investment organisations for the bank and the insurance company	0

F. *Marketing integration ➟ marketing synergies*
This is an average of the next three scores, based on:
F.a. The number of different channels used:

1.	(General) bank	1
2.	Savings bank, mortgage bank, or bank working with tied agents	1
3.	Insurance brokers	1
4.	Tied agents	1
5.	Direct channel	1

F.b. Product development process:

1.	New integrated products are made (cfr. a modular concept, with both banking and insurance elements)	4
2.	Package of banking and insurance products (bundling approach)	3
3.	Assortment policy (gearing of the production activities)	1
4.	No co-operation in the product development process	0

F.c. Corporate communication:
There is a corporate name and it is used for (some of) the subsidiaries

1.	Corporate name often used for the subsidiaries (bank/insurer) (in more than 90 %)	5
2.	The corporate name is frequently used (+ 50%)	3
3.	The corporate name is seldom used	1
	No corporate name	0

Appendix 2: Operationalisation of the corporate governance variable

1.	Structure companies	5
2.	Reference shareholders	4
3.	Family shareholders	3
4.	Subsidiaries of foreign groups	2
5.	Co-operations	1

Structure companies are a typical Dutch phenomenon (see Kabir e.a. (1997) for a detailed explanation) where the link between managers and stockholders is large, so here is much room for agency conflicts.

The reference shareholder is a Belgian phenomenon where a shareholder, without having the majority of capital, is able to influence the strategic decisions (because of the majority of the votes). However there can arise problems with other shareholders and sometimes this results in unstable and delicate governance structures.

With family shareholders some families have the shares and are represented in the management board.

As subsidiaries of foreign groups are under full control of stockholders, managers have not much room to pursue their own goals. There is some information asymmetry (the managers should be better aware of the domestic situation than the stockholders), but this asymmetry is restricted.

We expect that stockholders of co-operations have relatively more means to control managers adequately. Co-operations have emerged in the same way as institutions which share a common idea or a common characteristic (e.g. co-operations based on political conviction). Normally representatives of this larger co-operation have a seat in the board of the financial institutions of this co-operation.

PART C

Creating the future with all finance and financial conglomerates

7

THE BASIC CONCEPT OF CREATING THE FUTURE: WHAT STRATEGIC CONCEPTS AND INSIGHTS ARE USEFUL OR EVEN NECESSARY?

INTRODUCTION

Scientific literature as well as business practice show an increasing interest in strategic thinking and strategic policy making. It is hard to cope with the frequency of new publications in this area. This is especially true for business practitioners, who are not always able to apply the new ideas and new recipes as such in their daily business. Due to the hard competitive environment, financial and insurance companies have to struggle in order to survive and hopefully stay successful. It is therefore necessary that strategic thinking and strategic orientation are developed. It seems interesting to verify which useful lessons can be drawn from the new strategic concepts and to what extent these strategic recipes can or even have to be applied by banks and insurance companies.

Part of this analysis is based on a practical research done in the Netherlands with some of the larger financial conglomerates as well as some of the more specialised market players in both the banking and the insurance business.

ATTACK IS THE BEST OF ALL DEFENCES

Most banks and insurance companies have lately been involved in either establishing new strategies, thinking of new strategic planning or setting up new strategic guidelines. This historically unknown interest in strategic thinking has of course much to do with the hectic environment in which financial institutions must compete today. Gone are the days of stable market environments, protected by comfortable and secure (self) regulations.

As long as a company can operate in a stable environment, it can exploit its market position and develop a long term planning by extrapolating the past experience. In a raplex[9] environment everything is changing so drastically and so frequently that the only way to develop a strategic policy is by looking forward instead of looking backward. The new strategic framework of Prahalad and Hamel (1994) written down in one of the most known management books of the 1990s -*Competing for the Future: Breakthrough strategies for seizing control of your industry and creating the markets of tomorrow*- brightly prescribes that business firms should anticipate with an offensive and innovative strategy instead of sticking to a defensive approach. Firms must concentrate more on changes (on differences, regenerating and genetic variations) instead of continuity (managers must learn to forget, must break the overwhelming reference board).

This is a very difficult task for a business sector which historically has always been defined as being very defensive and even lagging behind. Nevertheless the many innovations in the financial and insurance sector prove that modern managers have understood the message very clearly: to win the future battle, it is more important to be flexible and to be fast in anticipating, than it is important to have a strong lobby, to have a good defence and to secure the possessions acquired in the past. Not only business firms themselves, but also the trade associations of banks, insurance companies and brokers must be conscious of this turnaround.

WHICH ARE THE MOST IMPORTANT DISCONTINUITIES THAT COULD ALTER THE COMPETITIVE RULES IN THE FINANCIAL SECTOR?

One could answer this question by using the words of a manager of a financial conglomerate: "The financial sector will tremble on its grounds for quite a long time".

Within the scope of this contribution it is not possible to develop a comprehensive analysis of all potential discontinuities; nevertheless it is interesting to mention some of the most important factors:

- destruction of the market protection due to deregulation, prohibition of market/cartel agreements, global competition, etc.;
- demographic changes and their effects on the financial stability of the welfare state;
- impact of consumerism, new life styles, changing customers' needs;
- technological facilities on the level of administration and business organisation as well as for marketing and

[9] Raplex stands for both *rapid* and *complex* changes.

distribution purposes;

- the information and knowledge society where the power is shifting more and more towards the management of information flows and information systems;
- and last but not least, foreign invaders and non-traditional competitors.

Large, well-known firms are no longer challengers because they fear that their existing portfolio could be damaged too much through cannibalisation. Nevertheless, this cannibalisation danger does not give them a free ticket to ignore the unpleasant truth neither does it make regeneration unnecessary. In fact all firms will have to regenerate unless they want to vanish as happened with the dinosaurs. The new strategic gurus, Prahalad and Hamel, are rather pessimistic about the ability of insurers to cope with future opportunities and threats such as consumerism, deregulation and elimination of intermediaries. Therefore they suggest that insurers should analyse very carefully what is happening in other sectors.

Being involved in the insurance and banking sector on a business level as well as on a scientific one, we would like to alter this hard point of view to a certain extent. The insurance sector today presents a very heterogeneous picture: besides some defensive insurers, quite a number of offensive competitors prove that they use the new strategic recipes, be it not literally. Even the legal and actuarial disciplines, which are the base of the insurance technique, can no longer be limited to a historical perspective. Due to the many changes on the level of *risks* as well as on the level of *laws and regulation*, the insurance technique changes from a retro-active to a more prospective focus. Also on the level of marketing and information technology, the scope is much more innovative and forward looking than it has ever been.

FROM THE ACTUAL COMPETITION TOWARDS COMPETING FOR THE FUTURE

In a stable market environment, much interest is devoted to optimisation of the existing activities and the acquired market share. The new strategic thinking breaks a lance in order to replace the focus from the actual position (market share, actual activities) towards competing for the future. This hypothesis in fact parallels or continues the existing lines developed by the classical portfolio thinking. In these classical portfolio recipes it is suggested that a firm should always aim at a balanced portfolio of activities with sufficient attention to future developments and potential new markets (the so-called question marks). The new strategic thinking goes much further in that direction. It is pointed out that the actual market share is not a good basis because it is a historical position, a historical value which gives no guarantee at all for future success. In strong turbulent environments, managers can no longer be experts of the past, they must become experts of the future. It is therefore more important to aim at future leadership and at *future market share* and this with the help of the necessary competencies.

The time horizon for strategic developments is not the coming year, not even the coming 2 or 3 years, but the far future. Therefore persistence (triathlon) and consistency are much more important than short-term success (sprint). These statements are sometimes incompatible with the short term during which top managers are responsible for the development of their company. So, future success supposes that the organisation and its incentive structure stimulate persistence instead of rewarding short-term success.

Insurance companies have been oriented towards the long term much more than their banking partners. Proof of this statement is to be found, for instance, in the way the (technical) reserves are used as a buffer to equalise and stabilise the business results. Nevertheless, in the insurance as well as in the banking sector an increasing importance is given to short term performance especially by companies quoted on the stock exchange. Many signals point in this direction: think e.g. of the increasing use of defensive instruments for speculative purposes (see the world-wide problems due to the speculative use of derivatives) to put forward future benefits (see the use of embedded value techniques, not only for internal calculation purposes but also for external reporting), implicit underreserving in certain firms (see the risk of life and pension insurance in relation to the much longer life time). It is therefore not astonishing to observe that the regulatory authorities are pleading more and more for harder regulation to secure the long-term stability and security of a sector for which guarantee and security is a basic good. To our opinion there is however a real danger that this *wheel of regulation* goes a little bit too far thereby reducing creativity and entrepreneurship.

Therefore, the financial sector has a great deal of interest in taking its own responsibilities and in giving priority to the long-term strategy. This does not mean however that the managers can ignore the actual market struggle. Even on the contrary, in a very turbulent market environment, action will often have to prevail upon reflection. Nevertheless, long-term success will not be possible without a broader focus on competitive distinctiveness in the long term. To anticipate upon *real* improvements for the (potential) clients is of primary importance for the development of this future competitive distinctiveness. To what extent the client himself can play a role in this detection of potential improvements is a hot issue which will be highlighted in the next point.

WHAT IS THE ROLE OF THE CLIENT IN SHAPING THE FUTURE?

Prahalad and Hamel give the impression that they are no defenders of the basic methods of modern marketing strategy. To our opinion they do not value the client high enough, which can be derived from the many contradictory remarks they make in this respect.

They state that the client has no forward-looking opinion. They pretend that consumers do not know what they want; so it is up to the enterprise to guide customers and to educate them with the help of appropriate communication. Nevertheless, they agree that future success of the firm can only be based on real improvements for the client; research must be done to investigate many aspects of these potential improvements:

- what types of technology, product and service concepts are best suited to answer the needs of the clients;
- what channels of distribution have to be used;
- what specific characteristics is a client looking for in a certain product: a firm should aim at the development of activities the client appreciates most;
- another statement made by Prahalad and Hamel is that a firm should always verify if its core competence measures up to what the client needs.

All types of market research are denigrated as being not useful. Some of the statements made by Prahalad and Hamel in this direction are the following:

> "Market research directed towards new product concepts is unreliable" (Prahalad and Hamel, 1994: 142);
>
> "You can only get a thorough insight in the wishes of the consumer through direct contact with the consumer himself and not through market research done by an external partner" (ibid. 114);
>
> "All too often questions used in market research give no possibilities for fundamental changes of traditional product concepts or for the creation of a real competitive advantage" (ibid. 110);
>
> "Insights into the new possibilities can be gained in many ways, ways that go much further than the traditional methods of market research."

We do not agree with these statements because a fundamental, scientifically constructed, market research is the only possibility to investigate the necessary improvements for the client. Of course it is true that clients are not able themselves to tell you what types of products they need, but they certainly have the possibility of explaining you what they are looking for, what their real needs are and to what extent these needs are answered by the existing supply in the market (may be not fulfilled at all, or even in a wrong way). In this respect, the technical product-oriented line of thinking must naturally be changed into a more functional thinking which supposes special creativity and innovation. Here the insurance sector is a very bad example. Their challenge is to succeed in changing the technical policy attitude per risk category in an integrated client-oriented solution.

TO WHAT EXTENT CAN A FIRM ITSELF DIRECT ITS FUTURE CHANGES AND BE SUCCESSFUL IN THE FUTURE?

This is perhaps the territory where Prahalad and Hamel have been the most innovative and delivered the most pioneering research. The basic recipe of leadership of the future is built around 4 important ingredients: stretch, core competence, leverage and networking.

All these concepts are useful for strategic policy making in the banking and the insurance sector; of course they ask for certain specific translations so that these concepts can be used in real practice.

The concept of stretch

The concept of stretch means that the firm must aim at far reaching, ambitious goals rather than stick to the classical recipe by which ambitions have to be reduced in function of the limitations set by internal or external environmental factors. This classical strategic thinking aims at co-ordinating what the firm wants to do, what it can do given the external opportunities and threats and what it should do taking the internal strengths and weaknesses into account. Consequently ambitions are reduced by limitations found in the SWOT analysis, whereas Prahalad and Hamel suggest that any firm should have a strategic intent that is unbelievably ambitious and which forms an obsession for everyone in the company.

This way of thinking can easily be applied to the banking and insurance sector, we can refer to one of our publications (Van den Berghe, 1994). This stretch thinking leads to a very important consequence, namely that success is not in the first place realised by firms which are shrinking but with firms building up their empire. The big changes in strategic thinking can be referred to as denominator thinking versus numerator thinking. What is it all about? Well, efficiency, profitability can be measured by comparing output and input. The output can be referred to as the numerator, the input as the denominator. So the classical strategic recipe focuses on the efficiency increase through cost cutting strategies and is therefore referred to as the denominator thinking (decreasing input). The new strategic line of thinking developed by Prahalad and Hamel focuses much more on improving the efficiency by increasing the output instead of lowering the input (so numerator thinking is seen as the best solution). They refer to the overwhelming emphasis on productivity increase, cost cutting, business processing re-engineering, lean and mean strategies as a form of anorexia. According to their stretch thinking, a firm should first of all aim at improving its position by ambitious goals, by increasing its organisation in the direction of growth. Nevertheless this does not mean that Prahalad and Hamel are defenders of a growth as such, because it can lead to different disadvantages. If growth is such an overwhelming goal, it easily leads to unrelated activities or non-attractive market shares.

This is certainly a very important message for the banking and insurance sector that has to cope with price competition, which is often approached by hard cost cutting strategies. Also think of phenomena such as cash flow underwriting and dumping practices just to reach a certain market share or to get a certain market entrance. Certain firms have however already applied the new strategic ideas in the direction of numerator thinking: this can be proven very easily just by referring to the many diversifications in the direction of all finance, financial conglomerates, and so on.

The concept of core competence

This is the heart of the new strategic thinking: the competitive advantage is not so much a question of better material assets or a better financial power, but the future is only successful for those who have the necessary core competence. Core competence can be defined as a package of competencies and technologies which allows a company to give the client a certain advantage.

The main building blocs for the future are therefore knowledge, talent, technology, but also alliance partners, intellectual property rights, market orientation and total quality management. Since these core competencies exceed the classical borders of strategic business units, the real competition is between firms or conglomerates rather than between strategic business units or products. These are valuable ideas for the banking and insurance sector. The question however is how far competition on a product basis is not still overwhelmingly a fact, especially in a sector that operates with independent distribution channels, such as the insurance sector.

Core competence is hard to define in practice. To test potential core competencies, three essential conditions have to be fulfilled:

- it must deliver fundamental value for the customer;
- it must differentiate the firm from its competitors (competitive distinction will necessitate a certain form of benchmarking);
- it must be useful for other types of products, other fields of activity or for external distribution.

Especially interesting for the insurance sector is the reference made to the importance of the distribution policy. To win leadership, it is not possible that a firm should be limited by its own distribution channels, it must have the chance to learn and to use all the available alternatives. In the insurance and banking practice, core competencies are often defined in the direction of technological leadership and customer orientation. They attach less importance to the combination of core competence through networking and alliances; most of them clearly prefer the integration approach in the direction of financial conglomerates. Nevertheless, quite some successful co-operation and networking formulas can be observed e.g. to create an adequate infrastructure for the electronic banking traffic or to collaborate in cross-selling (cf. bancassurance or assurfinance).

The concept of leverage

This concept can be applied in different directions. Material capital goods have to be depreciated because they obsolete technically (through the use of it) and economically (they get outdated - e.g. computers). Core competence on the contrary improves and is upgraded by using it; this creates a positive leverage effect that can be found in the so-called learning environment. Consequently, the learning curve must not be explained by the volume or output produced but through its positive leverage effect.

To our opinion this concept lives much more in service companies than in industrial firms, because knowledge and ability are often the essential goods sold in service firms. Especially in insurance firms with a traditional distribution system, it is of the utmost importance to succeed in developing a learning environment in order to improve the efficiency of the service chain -the *servuction system*- and to lower the percentage of mistakes and double work. Core competencies do not fit into the boundaries of a specific business unit but have to be developed across the product and business unit boundaries. The allocation of scarce resources per business unit or product line is therefore much less relevant than the optimal improvement of the available resources.

Sticking to the business unit concept does not lead to sufficient leverage effect nor to the detection of new market opportunities. This general recipe is hardly applied in the insurance sector, where technical aspects and regulatory prescriptions have led to a product-oriented focus. Only trying to develop a more market-oriented segmentation and organisation is already seen as a kind of revolutionary change.

Diversification strategies should also be based on the specific core competencies that are available or necessary. Prahalad and Hamel are in fact no great defenders of diversification because they are very critical about the classical economies of scale and scope. Focus and convergence is the aim; diversification is only justified in the scope of a logical improvement or better use of core competencies. Actually the financial sector is witnessing a large and varied number of diversification initiatives. Although this diversification is referred to as a *financial conglomerate*, it is certainly not a question of non-related diversifications as it was the case with many industrial and mixed conglomerates. Nevertheless the message by Porter, reinforced by Prahalad and Hamel, that diversification can quickly become unsuccessful, is certainly worthwhile to remember for every financial firm looking for diversification opportunities.

A positive leverage effect can also be gained by making optimal use of the core competence of others. The establishment of networks and coalitions has led to great success for industrial firms (e.g. just-in-time application, outsourcing). Also the banking and the insurance sector is giving clear evidence of this trend:

- the outsourcing of investment management or of the financial back-office;
- the outsourcing of claim settlement by banks diversifying into bancassurance (e.g. ABN-AMRO in the Netherlands with Centraal Beheer and Kredietbank in Belgium with Omniver/Fidelitas);
- the collaboration between insurers and sickness funds in the Netherlands.

CRITICAL EVALUATION OF THE SCALE OF A FIRM

The strategic thinking of Prahalad and Hamel in relation to the optimal scale of a successful firm is rather confusing.

On the one hand they defend the hypothesis that small firms are often at the origin of the battle for the future (leadership). This hypothesis is proven by the fact that they are more flexible and offensive in anticipating and exploiting the future trends. Large firms on the contrary are pictured as bureaucratic, undermining creativity and innovation, and underestimating the market intruders. They are often arrogant and vegetate on their birth certificate. They often ignore that the fruit of success contains the seed for failure. However the more the borders of a business sector disappear, which is the case for the whole financial services industry, the more the birth certificate of a firm looses its importance.

These observations are very valuable for the financial and the insurance sector. Most of the innovations have been forced by free riders who did not follow the self-regulatory rules or tried to circumvent the official limitations. This offensive and innovative approach is sometimes necessary in order to gain a considerable market position. Moreover, such aggressive competition is possible because market intruders do not fear cannibalisation, because they often start from scratch. Nevertheless we do not agree completely in saying that the large financial firms are becoming dinosaurs. First of all they are often free riders themselves in penetrating foreign markets. Moreover some larger firms have developed quite an aggressive attack to secure their long-term success. It is often thanks to modern technology that even larger firms can become more flexible.

On the other hand, Prahalad and Hamel are not completely against all forms of economies of scale. *They favour large scale initiatives* in the field of global brand names and global distribution networks. Also the positive leverage effect of the learning organisation will probably be more important in larger firms (more education facilities etc.).

This opposite view is explained by the fact that the larger scale must not be realised per firm but more by forming alliances, joint ventures and making optimal use of outsourcing. This networking improves the positive leverage effect, while a global market penetration and a global distribution system can be developed without the disadvantages of bureaucratic systems and organisations, so typical for large firms. Small firms are therefore considered to be of great importance for the modernisation of the different markets. Nevertheless they will not be able to take over the future leadership on their own. The optimal solution is therefore to combine forces between the larger and the smaller firms. A larger scale delivers the necessary grounds for a successful exploitation of innovations while the smaller scale procures the necessary entrepreneurship and flexibility. Larger firms on their own quickly become dinosaurs while smaller firms alone can suffer from impotence.

The role of larger firms in capitalising on the spin-offs of smaller colleagues is as important for market success as the procurement of venture capital to stimulate entrepreneurship. The only warning that has to be made is that a successful firm must watch over the intellectual leadership of the most crucial parts of the value chain in order to secure its competitive distinctiveness and to prevent that the dependence on other firms becomes too important (e.g. in relation to product development, distribution, etc.).

Many financial firms do not agree -for the moment- that networking is the best solution because they find it hard to realise potential synergies and to develop a long-lasting relationship. Many co-operation formulas failed (e.g. international banking networks) or were replaced by more integrated structures (e.g. global brokerage, global employee benefit structures).

CRITICAL EYE ON THE DISTRIBUTION

Global distribution is seen as a necessary condition for a successful exploitation of innovations and costly brand names. Prahalad and Hamel are very sceptical about firms that define their distribution method too narrowly, or that are afraid of experimenting in the alternative distribution methods. Very instructive, also for the financial and especially the insurance sector is the following comparison. A firm can loose a great deal when a new product or service with the dimensions of a fire-hose has to be distributed through a channel with the measures of a gardening-hose.

They explicitly refer to the distribution of insurance through agents and brokers: these classical distribution outlets are considered to have no future at all. Probably Prahalad and Hamel underestimate the entrepreneurship and flexibility of the independent firms working in this field. Of course, this does not mean that many new alternatives will not be developed and exploited further, such as direct selling, electronic selling, bancassurance and partnership formulas with third parties (such as e.g. automobile factories, affinity groups, employers, etc.).

DEFINE YOUR OWN FUTURE

Through the optimal use of the basic concepts of stretch, leverage, core competencies and networking a firm can influence its own future, it can alter the boundaries and the definition of the branch, the sector or the market in which it is working, or break the rules instead of following them.

A firm must therefore devote a great deal of importance to the definition of its core business, not in technical terms but from a functional perspective. This is especially important in the banking and the insurance sector where the future role can be completely different than it is today. Think e.g. of the intermediary function performed by banks and insurers in relation to larger clients; to what extent can the electronic highway decrease the information gap (and information costs) upon which this intermediary role is based? To what extent will industrial clients take over their role, as has already been the case with in-house banking and the establishment of captives? To what extent will the boundaries between banking and insurance institutions and products further be blurred and perhaps disappear in certain respects? Will insurers become the suppliers of material and technical support facilities rather than or besides their risk spreading and risk financing role? How can the destruction of the welfare state increase the territory of insurers?

8

CASE STUDIES OF STRETCH AND LEVERAGE

In chapter 6 we have shown that diversification in se did not reveal positive results, even rather the opposite. However we see that more and more groups try to complement their banking business with insurance activities and vice versa. At least it must be said that the *most famous* initiatives and examples perform well.

The purpose of this chapter is to look at three groups where *all finance* is at the core of their strategy. In a last paragraph we want to indicate that a creative vision on the synergies between (investment) banking and (re)insurance can also cause drastic shifts in the whole financial services industry.

THE ING CASE

The origin of the ING Group

ING Group formerly known as Internationale Nederlanden Group was the result from the merger in March 1991 between the insurance group Nationale-Nederlanden (NN) and the banking group NMB-Postbank Group. This latter group was itself the result of a merger between two important Dutch banks in October 1989. This merger between the largest insurer and the third largest banking group in the Netherlands attracted a lot of attention, however, not always positive.

There has been a huge discussion with Aegon, one of the larger shareholders of NN and at the same time one of its largest competitors, who claimed that the combination of a bank and an insurer was at the disadvantage of the insurer. As financial institutions are strongly dependent upon the capital market for their further expansion, it is extremely important to have a good price/earnings ratio. However this ratio differs considerably for banks and insurance companies. At the time of the merger: the price/earnings ratio for Dutch insurance companies was on the average 9 to 10, for Dutch banks this ratio was only 5 to 6. And as the newly created combination was measured as a weighted combination of a bank and an insurer, this ratio dropped considerably for the shareholders of NN. Thus it was stated that the banking-with-insurance concept only created value for the banking part.

According to CEO Aad Jacobs, the price/earnings ratio of ING is still a problem, and he considers ING compared to other financial institutions to be undervalued. However if we look at the performance evolution of ING, we can say that ING has performed very well.

Figure 8.1 ING Group share price

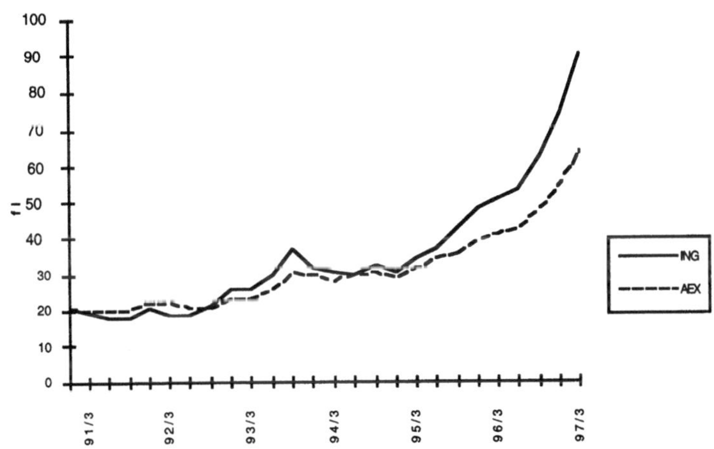

Although financial conglomerates have emerged at great speed, there have been hardly any cases where the integration between the bank and the insurance group has been stronger than in the case of ING. Indeed, integration has always been one of the main focus points of ING's strategy:

"Most publications speak about bancassurance or allfinanz, which, in most cases, refers to selling insurance via banking distribution channels or the other way around. Right from the start, ING's objective has stretched beyond such cross-selling. Our aim is to provide a complete range of financial services to both personal and corporate clients through the distribution channel of their choice [...] While many European financial institutions decided on looser forms of co-operation, Nationale-Nederlanden, Holland's largest insurer, and NMB Postbank Group, Holland's third largest bank, opted for a full merger in 1991" (A. Jacobs, speech at the Financial Services Industry Association, May 1995, Dublin).

"We aimed at a full merger, because in our opinion one strategic vision and one management structure would create the environment for future growth and achieving an integrated financial services approach. It would mean total commitment and focus on an integrated financial services strategy and marketing, instead of just commercial agreement for the home market, whereas such an agreement is by nature restricted in time and scope. A merger would also mean greater continuity and more flexibility with regard to market changes than a joint venture would have meant" (J. Holsboer, Geneva Lecture, September 1993, Rotterdam).

A new organisation structure: structure follows strategy

ING has always considered banking and insurance to be part of the same market for financial services and considers the traditional separation between banking, investment and insurance out-of-time. Moreover they believe that developments at the demand side create the need for a more integrated approach. This new approach was given shape by the creation of a new corporate structure in 1994.

Figure 8.2 ING's former organisation structure

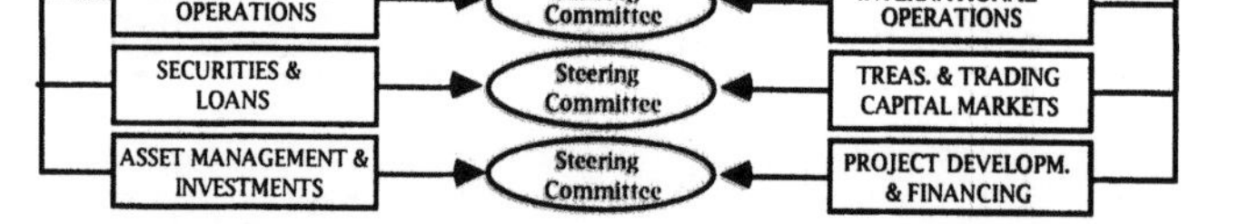

ING has kept the old organisation structure for three years. This period enabled them to get the cultures somewhat in the same direction. Synergy was achieved by creating steering committees, i.e. committees who investigated where and how the most important synergies could be exploited.

But the management of ING felt that in order to create a more flexible and integrated group, fundamental changes were necessary. In 1994 a number of such strategic changes were carried out in order to have better control over the activities of this diversified international financial services group. One of the most important decisions was the change of the management structure with one Executive Board. In this way the management served as the most notorious example of the group's integration strategy. Another major decision was to create a new organisation structure, which goes beyond the traditional separation within the financial services industry. The following business units were created:

- ING Nederland, which comprises the financial services to personal and corporate customers in the Netherlands;
- ING Financial Services International, which encompasses all the insurance activities outside the Netherlands in combination with the banking activities for private customers and small and medium-sized enterprises in local markets;
- ING Corporate & Capital Markets, which includes the operations in the money and capital markets as well as the international banking activities for corporate clients;
- ING Asset Management, which acts as the property and the relation manager (both for the clients as well as for ING itself): in this way it unites the asset management, real estate and private equity operations (ING Annual report, 1996).

Figure 8.3 ING's new organisation structure (since 1994)

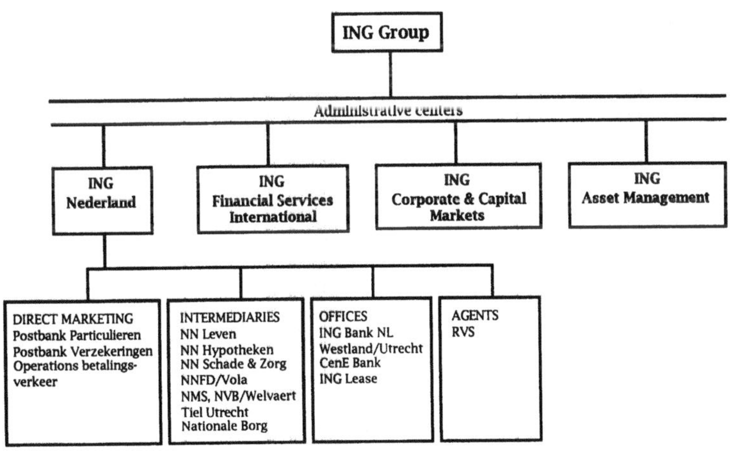

According to A. Jacobs, the integration has demanded much attention and energy because the cultures of the bank and the insurer sometimes clashed. In order to minimalise these conflicts great emphasis was placed on fostering one ING culture in which the best elements of the banking and the insurance cultures are combined. Furthermore the people should be made aware to think beyond the borders of their own business unit and look at what is best for the Group as a whole. Effective internal communication plays a key role in this process. This is not always easy. Everybody considers synergy as the *1 + 1 = 3 effect*. But this picture is somewhat irrealistic: it is rather a *1.1* + *1.2* which sums to *2.3* (compared to the original *1* + *1* = 2). In this case no problems occur because each business unit benefits from co-operation and integration. This situation changes when it becomes *0.9* + *1.2* = *2.1*. This is synergy as well: it is clear that the Group as a whole benefits from co-operation between the two units. ING has experienced a comparable situation -even in the beginning of its existence- when they announced to sell insurance products via the Postbank. It is worth noting that the situation explained below in the box occurred after the announcement of the merger, and thus long before this new organisation structure was created.

The boycott of ING's independent insurance intermediaries

As the Netherlands' largest insurance group, NN had developed close and harmonious relations with its most important distribution channel, the independent insurance intermediaries that are responsible for approximately 85 percent of NN's product distribution. There seemed to be little indication that NN was prepared to disturb that relationship. In an interview given in July 1990, Sam Jonker, member of the management board at NN, was quick to point out the need to ensure that the use of alternative distribution channels, such as banks, did not upset the equilibrium as far as its insurance brokers were concerned.

"I think banks and insurance companies have to be very careful about the way they treat each other [...] A company like Nationale-Nederlanden has to watch that it keeps a very good relationship with the independent agents that are very important to us [...] If you try to marry these two cultures together, you have to do it in a very careful way that is acceptable to your agents" (Interview appeared in 'Life Insurance International', Issue No 18, September 1990, Lafferty Publications).

However, just a few months later, NN was announcing that it planned to abandon its independent status in order to merge with NMB Postbank Group. The proposal caused an immediate upheaval among a number of interested groups. What caused the most critique was NN's intention to introduce specially designed insurance products to be sold through the Postbank's large client base of over 5 million accounts. With lower distribution costs, these products would inevitably be cheaper than those offered through NN's brokers.

Though the insurer hastened to assure its major distribution channel that it had no intention of trying to replace it with the NMB Postbank network, the brokers' response to the introduction of dual pricing was immediate and hard. At the annual meeting of the brokers' main organisation, the Nederlandse Vereniging van Makelaars in Assurantiën en Assurantieadviseurs (NVA) held on November the 22, 1990 , the vast majority of delegates voted to impose what was effectively a boycott of NN products. Under the ban, brokers agreed to do no new business with NN, unless it was specifically in their client's interest [...]

However in September 1991, the NVA announced that they were lifting their boycott of NN products, following a promise by the insurer that he would only sell his simplest policies, such as travel insurance and certain types of annuities, through the Postbank. NN agreed to continue selling car, fire and liability insurance via the brokers.

(*In* Lafferty Business Research, *The Allfinanz Revolution: Winning strategies for the 1990s*, 1991, Lafferty Publications, Dublin.)

ING's core competences

Despite some early problems (e.g. the brokers' boycott) ING is very successful with its strategy of integration of banking and insurance business. Although the original objectives of all finance had to be adjusted, it seems that ING has found an optimal way of delivering integrated financial services. In this respect the Dutch home market serves as some kind of an all finance laboratory. It is important to notice that ING has a financial relation with 75 per cent of the Dutch households. In this way the Group has an enormous potential to cross-sell products of the different business units. For example, in 1995 ING Bank earned fl. 67 million on commissions of selling insurance products (1994: fl. 52.2 mln). And the gross premium income of life insurance through the Postbank has amounted to more than fl. 320 million. In spite of this, some analysts consider the commercial benefits in ING's home market relatively modest. Synergies are also achieved with the asset management and the corporate and investment banking department.

ING's home market is very important for the Group, but it has considerable business abroad. ING's strategy seeks to maintain its positions in developed markets and where possible extend them through organic growth and complementary acquisitions or participations. There too, they want to develop the all finance concept. Because banking was predominantly corporate while insurance was retail the possibilities looked limited a few years ago. However this is changing and more and more banking and insurance partners within the ING Group are looking for mutual synergies. As we already said, the experiences of the home market serve as some kind of a knowledge base which can be used by other business units abroad. The implementation of an all finance strategy can in other words be considered as a real core competence as defined in the new strategic management literature (see chapter 7).

Besides the all finance competence, ING is also a leader in emerging markets. According to the 1996 annual report, ING intends to strengthen its position in the emerging markets by investing in corporate and investment banking, asset management, (life) insurance and commercial banking. ING has built up significant know how in this respect and here again we can consider this as one of the core competences of the Group. Illustrative in this respect is a quote out of Global Finance (December, 1994: 78):

> "Nowadays, *emerging markets* is virtually a household term, with every global financial institution promoting its expertise in the area. But none has displayed the single-minded approach of ING, whose history in the business dates back to its key role in establishing the secondary market for LDC (less developed countries) debt during the early 1980s. And few have bet more on the future success of these unstable, often marginal economies than has this huge financial conglomerate that was created by a series of mergers in recent years [...] ING realizes it must dedicate a greater effort to becoming an investment banking power in emerging markets. Its insurance link will help, in part because ING Bank now has access to an in-house group of institutional investors. ING has already placed portions of several emerging market deals with its insurance arm."

Furthermore there is interaction with the all finance strategy because wherever possible, greenfield expansion is to be on the *full service* concept.

Of course the Group's international strategy was considerably reinforced by the acquisition of Barings. Barings and ING seemed to complement each other perfectly in the emerging markets. The asset management division reports to ING Asset Management while the other parts have been put under the Corporate & Investment Banking umbrella. Although the British investment bank remains quite autonomous, efforts are made to intensify co-operation with the ING Bank.

On the whole we can say that the future looks good for ING. While it has built a strong position in its home market serving all kinds of customers with all kind of distribution channels, the Group has also established a clear strategy to enhance in its core markets and to expand in emerging markets. Furthermore there is also a strong financial discipline for all subsidiaries which has resulted in a steady record of earnings growth.

THE FORTIS CASE

The origin of Fortis

Fortis is one of the largest competitors of ING in the Benelux market. This Group also is an adept of an integrated financial services approach and they have developed a clear strategy in this respect. The Fortis Group was founded in December 1990 when the Dutch AMEV-VSB Group decided to merge with the Belgian leading insurance group Assurances Générales (AG) and has evolved to an international insurance and banking group in Western Europe and the United States. The Group is also growing in Australia and Asia although on a relatively modest scale.

Fortis is a somewhat special case because there exists no legal entity Fortis. Fortis is the name for the operating subsidiaries of the two mother companies: the Belgian AG Group and the Dutch AMEV Group. These parent companies have retained their own independence and identity in full for several reasons: differences in legal structures, tax positions and differences in power exercised by the main shareholders. This structure was also used by other financial institutions aiming at cross-border alliances such as Crédit Local de France and Gemeentekrediet (1996) and recently by Zürich and the insurance companies of the British BAT (1997).

As we already explained in other chapters, both groups are financial conglomerates themselves. The AG Group had a very strong position in its home market and a solid financial position but relatively few international presence or banking activities. NV AMEV was a strong player on the Dutch insurance market with a relatively wide international radius, however, without playing a significant role in the main international markets. VSB Group, a bank with a solid financial position, originally a savings institution, had a reasonable strong position in the Dutch retail market but lacked national distribution for which it was seeking support.

The main companies of the Fortis Group are:

- Belgium:
  - AG 1824 (with Metropolitan, Krediet aan de Nijverheid)
  - ASLK-Verzekeringen
  - ASLK-Bank
- The Netherlands:
  - AMEV Nederland (with life, non-life, health, Falcon, Interlloyd, etc.)
  - VSB Group (with CDK Bank, TOP Lease, GWK Bank)
  - MeesPierson
- United States:
  - American Security Group
  - Fortis Financial Group
- International: different business units.

The merger between the Belgian and the Dutch financial groups has resulted in a remarkable financial conglomerate. Rather than being a target of (hostile) takeovers the newly created combination has acquired a substantial number of financial institutions and in this way has achieved substantial growth. In 1993 Fortis acquired the ASLK-Group which disposed of a large savings bank and an insurance company and thus strengthened its position in the Belgian home market. In 1992 Fortis created a joint venture, Caifor, with the largest Spanish savings bank La Caixa. This joint venture is very successful because this insurance group has become one of the four largest insurers in Spain. At the end of 1996 Fortis took over MeesPierson, the investment subsidiary of ABN-AMRO and has also strengthened its banking position in the Netherlands significantly.

Figure 8.4 Legal structure of the Fortis Group

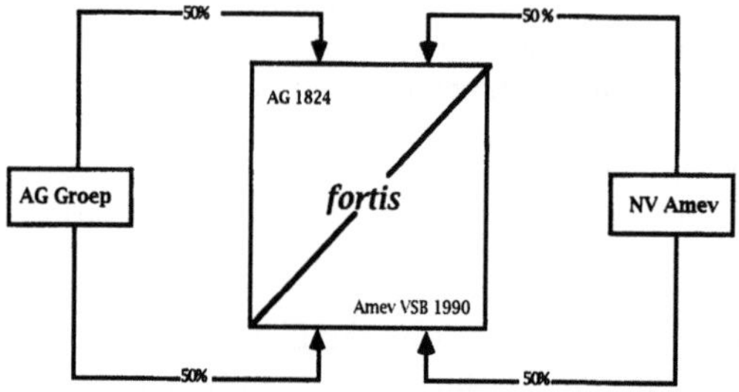

The strategy of Fortis

Fortis's strategy can be summarised in the following words: multi-domestic, multi-channel and multi-product. This strategy bears close resemblance to ING's strategy. Indeed Fortis has undertaken far-reaching decisions in order to support this strategy. Although it seems that Fortis consists of all different elements, Fortis is clearly a single entity, with a single management, a single strategy and a single group equity. According to Hans Bartelds, chairman of the Executive Board (together with Maurice Lippens), what counts is managing the differences.

> "This market strategy is hardly compatible with centralised management. However, it does require coherent action. This is why the regions and operating companies have *controlled freedom*. Controlled in terms of basic financial requirements, internal auditing and the pursuit of the Fortis strategy. But with total freedom to operate and present themselves in a manner which is appropriate to the local market. Of course in a manner which is everywhere distinctively Fortis: customer- and market-driven and with a will to supply distinctive, innovative services which exceed the customer's expectations. These shared values are an important element linking the Fortis companies" (Fortis Annual Report, 1995: 16).

The complex organisational structure has been recently changed towards the structure presented in figure 8.5. Four new operational holdings were created at the end of 1996, based on the geographical location. These four management centres are responsible for both the banking and insurance activities in the respective countries. Fortis International supports the management of the Fortis companies in seven countries and ensures the implementation of the international strategy.

Although this structure seems very decentralised there is controlled freedom which is achieved by a *linking pin* principle. The most important management committee is the Management Board responsible for the development and the operationalisation of the Group's strategy. This Management Board consists of 4 people:

the two CEOs of the mother companies and two vice-chairmen. These four members of the Management Board are also part of the Executive Board charged with the more operational control. This Executive Board consists of 15 people which are all responsible for the most important business units.

Figure 8.5 New organisation structure of Fortis (since 1996)

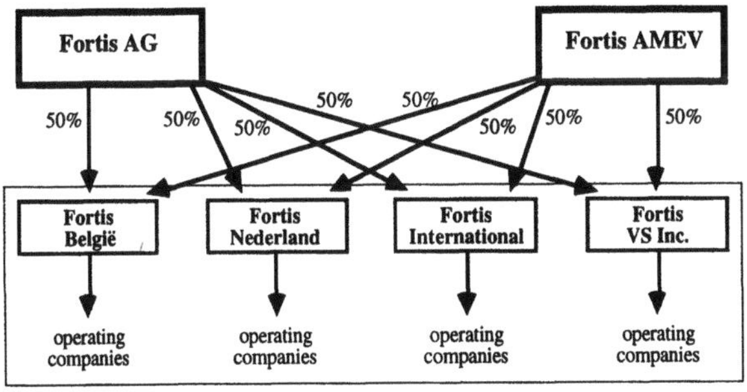

As we already mentioned each country is responsible for working out an all finance strategy which is dependent on the particular environment. Therefore Fortis Nederland has created a new organisation structure consisting of 7 business units each focusing on a particular product-market combination. What is remarkable is that the business unit structure has nothing to do with the legal structure of the former brands. For example three business units may make appeal to AMEV Leven to underwrite life risks.

In all other countries where Fortis is active, the legal structure and the business units run parallel. For example Fortis Belgium uses different brands to serve the independent intermediaries: AG 1824 to deliver insurance products, and Krediet aan de Nijverheid is used for banking products. In this way the assurfinance strategy is given shape; for the bancassurance strategy Fortis Belgium uses the bank branches of the ASLK-Bank. Insurance products are delivered by ASLK-Verzekeringen. Just like in the Netherlands, the investments are combined in a separate business unit: Fortis Investments Belgium. It shows that the Group -at least in Belgium and the Netherlands- focuses on distribution and investment synergies. It must however be mentioned that Fortis uses its investment department more in a traditional way and does not as that far as ING.

At least it is clear that one of the big challenges of the Group is to find an optimal equilibrium between the synergies created by centralisation and the flexibility of decentralisation. Hans Bartelds argues that there is room for *different* organisations.

"It is a question of managing the differences, not defensively but deliberately offensively so that the entire organisation can benefit from the strength of the differences [...] The bigger the organisation therefore becomes, the more room it must create for the diversity needed within its own ranks. This can only be achieved if it accepts that it does not need to organise everything centrally and uniformly. This is not a disadvantage. Quite the opposite. Instead of avoiding the countless differences that are associated with diversity, the organisation should explicitly create room for them. For example as a catalyst for synergy" (H. Bartelds, "Managing the Differences", IIS, July 1996, Amsterdam).

That this strategy has not been without success is shown by the share price evolution of the two mother companies.

Figure 8.6 Fortis AG share price

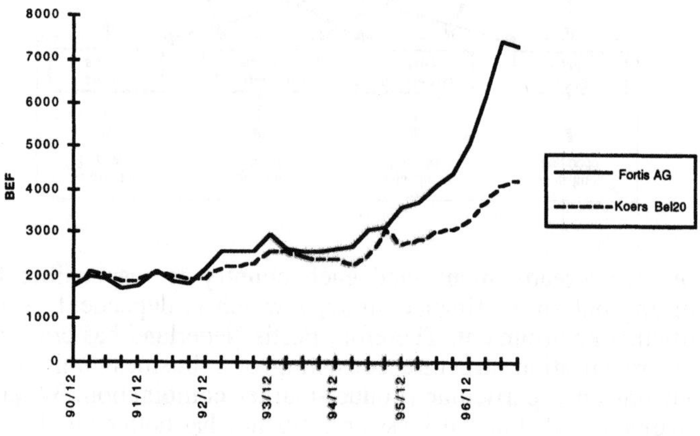

Figure 8.7 Fortis AMEV share price

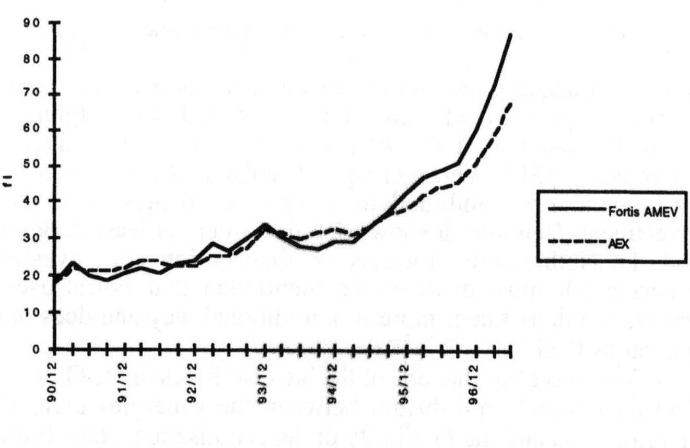

THE CASE BANCO COMERCIAL PORTUGUÊS (BCP)

BCP: the success story in Portugal

Banco Comercial Português (BCP) is one of the success stories in Portugal. This company became fully operative in 1986 and since then has become one of the largest financial services companies in the country. In 1995 it was able to acquire Banco Português do Atlântico, at that time the second largest Portuguese bank.

Table 8.1 Evolution of the BCP Group

	1987	1989	1991	1993	1995*
Total assets	129.746	519.223	1.206.324	1.927.158	5.404.395
Total deposits	105.102	404.384	972.140	1.532.970	4.381.151
ROE (%)	25.1	21.3	16.9	14.3	12.3
Market capitalisation (at year-end)	54.786	158.100	163.759	291.655	223.211

**: the figures of the BPA Group included*

Incorporated in June 1985 BCP established a clearly defined mission:

> "To become a major Portuguese Bank by providing, with excellent service, innovative and personalised products and services designed to fulfil all financial needs and expectations of the most relevant domestic market segments. The development and strategies of the BCP Financial Group are based on the need to cater to the international financial requirements of its customers and the commitment to a one-stop banking approach to the marketplace. Additionally it is seen that benefit may arise from the complementary business opportunities between a family of financial institutions."

Since the early days of the bank's existence the purpose has been to provide global financial services. At that time BCP had some minority participations in a number of specialised financial institutions which they expanded to majority stakes in order to benefit from synergies and complementary business. These minority stakes were purely related to the developing phase and in that stage BCP did not have the financial means. However where they could they tried to achieve control in the provision of the services. Thus as far as the market and delivery and customer relations concerned, BCP was the controlling party and because of their strong image they wanted to manage the relationship with the client.

Figure 8.8 Legal structure of the BCP Group (before the merger with BPA)

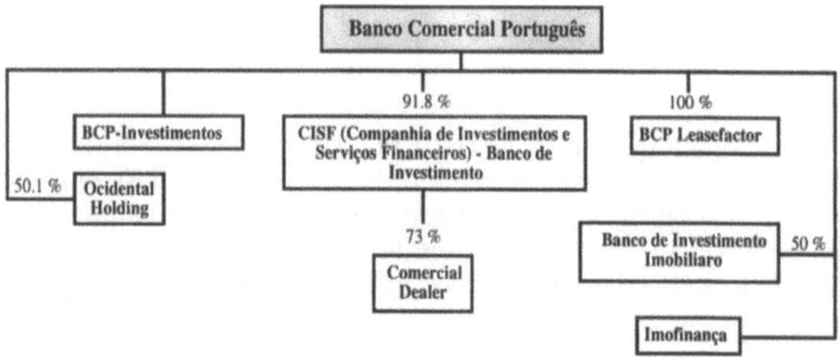

Their most important business units are scheduled in figure 8.8. The Group has considered a broad range of financial products. As the Group became larger and larger additional holdings were set up in order to have more adequate control. For example BCP Investimentos is a holding company which consists of a number of small companies which started from scratch and which are active in different related fields. CISF was a merchant which BCP transformed to an investment bank in 1992. Some subsidiaries of CISF were transferred to BCP Leasefactor, which is in turn a holding company comprising all leasing and factoring activities.

Also important with respect to the discussion on financial conglomerates was the insurance group Ocidental. It was strategically integrated into the bank in 1989 and the establishment of Ocidental Holding in April 1990 allowed BCP to combine all of its insurance interests under a co-ordinated management and financial structure. This move was intended to stimulate growth and cross-fertilisation and to build a strong insurance arm integrated into the overall development of the BCP Group.

Of course the acquisition of Banco Português do Atlântico (BPA) has had serious consequences for the strategy of the Group.

The BCP/BPA Group

In the wake of the take-over bid launched jointly with Companhia de Seguros Império, BCP acquired a 50.0001% holding in BPA. The selection of BPA as the institution best suited to the aim of strategic concentration was based on its tradition and notoriety, good geographical coverage of its branch network, loyalty of its customer base, solidity and quality of its balance sheet and considerable potential for an increase in its degree of involvement with its customers.

Following the acquisition of BPA, alterations were made to the shareholder base of the various subsidiary finance companies, which led to both BCP and BPA taking equal shareholdings in the various subsidiaries engaged in insurance, merchant banking, specialised finance and asset management activities. The business areas were incorporated into integrated companies in order to get the maximum advantage of the complementary synergies and competences, and to provide interlinkage of the strategic division and management of their businesses, without sacrificing each company's brand and distribution channels.

The restructuring of the BCP/BPA Group holdings into subsidiary companies involved: the acquisition by BPA, through a wholly-owned holding company, of 50% of the share capital of Banco CISF; the consolidation of all the direct and indirect holdings in insurance companies into Seguros e Pensões, Gere; the acquisition by BCP Investimentos of the investment fund management companies of the BPA Group; the consolidation of the BPA Group leasing and factoring companies into BCP Leasefactor under the new name of Leasefactor; and the integration of BPA Credinova into the structure of CrédiBanco.

The strategy of internationalisation involved the launch of three banks in strategically important markets, in order to make full use of the potential offered by the specific competitive advantages of the BCP/BPA Group.
(Annual Report, 1995: 41).

The key success factors of the BCP Group

Three elements have been at the heart of BCP's success until 1995.

Segmenting the market While most of the Portuguese banks used a standard approach and considered all consumers as grey mice, BCP has strongly focused towards different segments of the financial market. Initially BCP targeted affluent individuals and medium-sized companies which were given a superior service through the combination of traditional banking and the use of specific and modern systems made possible by the latest information technology.

Later on, new segments were added so that BCP's market segmentation could be presented in the following way.

Figure 8.9 Market segmentation of the BCP Group

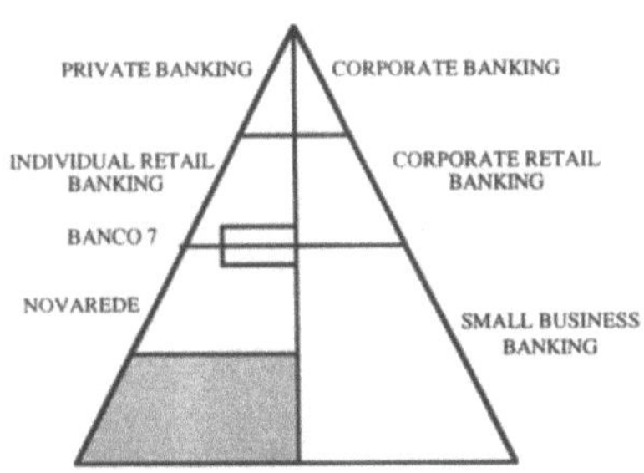

This concept of segmentation of the customer is the basis for the whole BCP philosophy. This figure of the triangle is driven towards everything they do: identifying what the customer wants and providing all these services to the clients in a fully integrated and co-ordinated way.

> "BCP continued to enhance its innovative and distinctive commercial strategy, based on the development of autonomous distribution networks offering different standards and a complete range of financial products and services designed to meet the specific requirements of each market segment."

- Individual retail banking: delivers financial products to affluent and high net worth individuals through a specialised approach and personalised relationships;
- Corporate retail banking: is catering for the needs of medium-sized companies
- Corporate banking: is approaching larger corporations;
- Private banking: addressing individuals with substantial financial portfolios;
- NovaRede: is suited to the financial needs of medium income individuals and small businesses;
- Small business banking: is aimed at the small companies and individual entrepreneurs that has a high business potential for deposit-taking, lending and complementary services;
- Banco 7: is the first and at least for the moment the only second-generation bank chiefly directed at the upper professional levels of the Portuguese population (which is more receptive to technological innovation and less vulnerable to adverse conditions in the employment market).

This market segmentation is used throughout the whole Group: for example there is a NovaRede manager in the insurance company who is responsible for determining what the needs are, for pricing the products, supplying the products and the service, etc. This segmentation triangle is also used in the board: while a manager is responsible for a certain product group, the same manager may also be responsible for a certain segment as well. Thus there is some form of matrix responsibility as well.

It is interesting to point out that for each market segment the Group has figures, not only on profitability but also on satisfaction by clients and employees. This is mainly obtained from the different management information systems the Group has developed. Furthermore, the Group has invested a lot in technology in order to create an integrated infrastructure which provides the base for sustained efficiency and innovation.

Building human resources "We are starting today. We have hundred years of experience" was the headline of the press announcements in May 1986 on the opening of BCP. This statement indicated that the bank disposed of an experienced team of professionals which were complemented by graduates of high schools and universities.

There are training courses for the employees and in 1992 the *Quality Improvement Process* was started up. The purpose of this process was to increase not only external customer satisfaction but also internal customer and employee satisfaction.

This was also confirmed by Jeff Medlock, manager of Eureko, the co-operation agreement between several European insurers (including Seguros e Pensões Gere, the current holding company for all insurance operations), who stated that it is a high quality Group. It is one of the most professional companies and they have a clear philosophy that everybody knows: it does not matter to whom you are talking, they all know what the area they are working in, is trying to do for the Group.

Building alliances The Group is quite unique in creating partnerships with other (financial) groups. The most notable example is the alliance with the Eureko Group. As already mentioned Eureko is a co-operation agreement between different insurance groups. Eureko combines the foreign subsidiaries of these insurance groups but leaves control in the respective home markets. Eureko has a 49.99 per cent stake in the insurance holding of the Group, which indicates that they are very important: so that they are sure that they will get full commitment.

There are partnerships with other financial institutions as well. More particularly the internationalisation strategy of the Group is based on a joint approach with partners experienced for their capacity to take advantage of the business potential in specific segments. For example there was the joint venture in 1992 with Banco Popular Espanõl -Banco Popular Comercial- designated to serve Spanish and Portuguese customers in France. In 1995 they set up Banco Internacional de Moçambique with joint efforts of the authorities and public entities in Mozambique. Of course the main ideas of the BCP Group will be transferred to the new bank so that here again some competence for the BCP Group becomes available.

INSURANCE DERIVATIVES AND SECURITISATION

Until now the focus has been on the interactions between banks and insurance companies on the retail market. Some groups have also indicated that they look for synergies in the corporate market but these are rather exceptions.

Nowadays we see that banks and insurance companies can benefit from each others' capabilities in totally different ways than what was originally expected. More precisely there are links emerging between reinsurance companies and investment bankers on the level of insurance derivatives and securitisation. This new phenomenon indicates that considering the banking and insurance business *creatively* opens up new possibilities for exploiting synergies !

What is it all about?

As the number and the size of natural catastrophes have significantly increased, insurance and reinsurance companies faced severe problems. The idea was even raised that the traditional insurance system was insufficient and that it faced (and still faces) a serious capacity gap. It is this capacity problem that forced the (re)insurers to look for other risk transfer methods.

In contrast, the capitalisation of the international financial markets is very large and its daily fluctuation -about 70 basis points or $133 billion on average- exceeds the maximum possible insurance loss which might arise from an earthquake catastrophe (Sigma No. 5, 1996).

According to P. Shimpi (1997) these new developments are part of a more general trend, namely the convergence of financial and insurance markets:

> "Finance, as practised by banks, has evolved over the years to encompass a significant level of risk management. Banks participate in transactions that expose them to various financial risks related to interest rates, foreign exchange rates, stock markets, commodities and credit, to name a few. These are managed using two fundamental risk management techniques, namely diversification and replication [...] Insurance risk management has not changed fundamentally since it started. The driving principle is diversification. It is a powerful tool. However, market forces are directing the attention of the insurance industry to the capital markets. Some efforts are underway to develop replication techniques for insurance risks" (p. 17-18).

Alternative methods of risk transformation: a new asset class?

The main characteristic of these alternative methods of risk is that the insurance risks are not traded within the traditional (re)insurance markets but are transferred to the much more liquid financial market. According to the aforementioned Swiss Re publication the first attempt to do this was made by the Chicago Board of Trade (CBOT) which has been trading futures on catastophe loss indices and related options since the end of 1992.

Generally a distinction can be made between (1) insurance derivatives and (2) catastrophe securities. We will briefly describe these new methods without going into detail.

Derivatives are financial instruments whereby the value is based on an underlying asset. For insurance derivatives this is a loss index which is provided daily by Property Claims Services (PCS), one of the leading organisations in the USA which estimates insured property/casualty losses. Options and futures are traded based on this index. This is illustrated clearly by Kielholz and Durrer (1997).

Trading in *PCS Cat Insurance Options* takes place mainly within the framework of so-called *call spreads* which are organised similarly to layers of traditional non-proportional reinsurance. This layered cover is achieved through a combination of two opposite positions: by simultaneously buying and selling call options with different strike values. The lower value represents the attachment point, whereas the upper value defines the exit point.

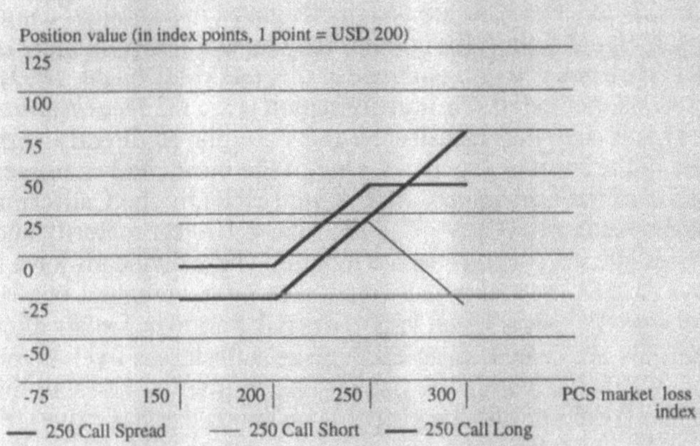

For example, an insurer wants to reduce its earthquake risk exposure on the American West Coast. It thus buys a number (adequate for its portfolio) of *200/250 Western Annual Call Spreads. The buying of 200 call options*, the first element of this double transaction, determines the priority of the cover: this begins as soon as the insured total loss in the Western region exceeds USD 200 billion (each index points represents an aggregated market loss of USD 100 million in the respective region). Through the *simultaneous sale of 250 call options*, the acquired reinsurance protection is limited to the area of market losses between USD 20 billion and 25 billion. In this way, the insurer can tailor catastrophe cover more precisely to its needs and buy this cover at a lower net option premium.
(Swiss Re, Sigma No. 5/1996: 10).

In this way derivates restrict the damage caused by a large natural catastrophe. There are however some problems with these derivatives. Until recently the information aspect has been underdeveloped so that investors had almost no idea of the consequences of their decisions. However the use of a new data pool by the CBOT (Chicago Board of Trade) has increased transparency and credibility so that confidence in the alternative hedging instruments is remarkably strengthened.

Another problem is that these options are rather standardised, as opposed to traditional reinsurance which is a tailored product (Smith e.a., 1997). Provided that the correlation between claims of an individual insurer and the movements in the index are close to 1, these options deliver an adequate coverage. Therefore these insurance derivatives are not substitutes for custom-made reinsurance programs, but are rather a complementary instrument.

In addition to the Chicago Board of Trade, an additional alternative is implemented in New York. CATEX is the name for this Catastrophe Risk Exchange which started to operate in October 1996. The difference with the CBOT is that there is no direct flow of additional capacity but the increase in available total capacity occurs indirectly. The risks are weighted against one another: a unit of exposure in one part of the country may be worth several units in another if someone is willing to trade at that rate (McDonald, 1997).

The second method of risk transformation is via the *securitisation model*. In this way the catastrophe risks are placed directly with investors in the form of securities. One of the most cited examples is the issue of the *Earthquake Risk Bond* (ERB) by the California Earthquake Authority (CEA). The ERB has a 10-year maturity and it generates interest coupons each 6 months. These dividends have a higher yield than the above risk-free return on government bonds. But of course the risk is much larger: the risk period is 4 years after the securities are issued. If an earthquake with losses over $7 bn occurs in California within the risk period, an investor loses all the future interest coupons (the principal is guaranteed at par value). In this way during four years the risk of a catastrophe is for the account of the investor.

For the moment the success of these new instruments is quite restricted for some reasons. But as more and more investors gain experience with derivatives and securitisation they may be helpful solutions for the capacity constraints on the reinsurance market. The study of Swiss Re has shown that for investors the creation of these new category of assets is an interesting element.

> "The inclusion of catastrophe risks as a new investment alternative opens opportunities for even more efficient portfolio selection. The additional, almost completely uncorrelated diversification possiblity shifts the so-called *efficient frontiers* upward, within both national and international modelling frameworks. This means that investors can achieve a portfolio mix with lower risk for each level of expected return, and with a higher return for each level of risk".

For a graphical representation we refer to the Swiss Re study which was published in Sigma No. 5/1996.

Some real-life examples of securitisation

Although the examples are restricted in number and value, it is interesting to mention some of them. According to McDonald (1997) the first successful private completion of a catastrophe-based surplus note (worth $400 million) was realised in 1995 and was sold by Nationwide Group, an American insurer. In 1996 other deals followed with an offering of Arkwright Mutual Insurance Co. (worth $100 million) and a deal completed at the end of December by St Paul Re, the reinsurance arm of St Paul Companies ($68.5 million).

But the breakthrough came in June when USAA placed $477 million of one-year hurricane bonds through a private placement with the help of investment brokers Merill Lynch, Goldman Sachs and Lehman Bros. The bonds were issued by a special-purpose reinsurance company of USAA, Residential Reinsurance Ltd.

In January 1997 the European-based insurer Winterthur was the first European insurer to complete the first public offering of catastrophe-based bonds, worth $566.5 million. And Swiss Re, a Swiss reinsurance competitor of Winterthur, also engaged in this business (October 1997).

The general opinion is that more deals will follow as more and more investors will be convinced of the possibilities of this new market. However there are still some problems that have to be solved in order to exploit the full potential of this market.

SECTION II

MANAGING THE RISKS INVOLVED

REGULATORY ISSUES AND SOLUTIONS

9

THE MAIN RISKS AND POLICY CONSIDERATIONS

BRIEF OVERVIEW OF THE POTENTIAL RISKS INVOLVED

The increasing interface between banks, insurance companies and investment institutions confirms that this is more than a fashion or a temporary diversification trend. It may however not be forgotten that not everyone is convinced of the advantages involved.

Regulators watch the diversification of banks and insurance companies very closely because they fear potential threats which increase the business and societal risks. Illustrative in this respect is the report published by the Tripartite Group of Bank, Securities and Insurance Regulators. This group was formed early 1993 "to address a range of issues relating to the supervision of financial conglomerates" (Tripartite Group, 1995).

In the United States, the supervisors are keeping an eye on the European scene, but they have not made the decision yet to break down the barriers that still exist between the different parts in the financial system.

In the next paragraph we will describe the potential risks involved in financial conglomeration.

Risks of contagion

When financial conglomerates are formed, it is feared that additional risks could arise. Supervisors are very concerned with the risk of instability and insolvency. They think that, with the creation of large financial groups, problems in one part of the organisation could infect another (healthy) part of the group (risk of contagion). This risk is comparable to a certain extent to the risk of large financial exposures; however, also important psychological effects on the image, reputation and credibility can increase the risk involved. Up until today, no evidence is available concerning this risk exposure. It is however questionable to what extent the total risk of the conglomerate is larger than the sum of the individual risks involved. To the extent that conglomerates procure the possibility of a better risk spreading it could be the opposite, namely that by diversifying one can decrease the global risk exposure. Several examples can be given in this respect:

- a better spreading of the business risk can offset the impact of economic cycles, underwriting cycles, investment cycles, etc.;
- a better spreading of the portfolio could decrease the investment and credit risks;
- a better Asset Liability Management could decrease the duration, interest and currency mismatch.

The risk of double gearing

Another aspect, related to the problem of insolvency and instability, is the risk of double gearing. Supervisors are afraid that capital will be used several times to determining the adequate capital coverage for the group as well as for the different parts of the conglomerate. The supervisors are unanimous that this danger should be prevented. As a consequence of the increased co-ordination between the supervisory authorities of insurance and other financial firms, a new proposal for a European Directive on the supervision of insurance groups has been formulated in order to integrate the prohibition of double gearing into the insurance regulations, hereby aligning them with the existing rules in the banking sector (for more details see chapter 10).

The creation of opaque structures

Other problems could arise as well. It is argued that financial conglomerates could be set up in such a complex way that the structure becomes opaque. In the first place this could lead to problems for supervisors, who want to assess the real risks of the conglomerate or who want to control the activities in an effective way. An even more dubious practice could be feared when some activities are transferred to those entities which are subject to less prudential supervision or which are not subject to any control at all, thereby searching for the way of the least resistance. The opaque structure could even be used to hide doubtful activities.

Again, the growing interface between banks and insurance companies led to a combined regulation to cope with the above mentioned risks (the so-called *post-BCCI directive*).

External conflicts of interest

In addition to the opaque structure, there are potential conflicts of interest that could have negative influence on the external business environment, such as clients, investors, supervisors, etc. For example what will happen when a firm gets in trouble and the bank arm has a credit exposure, while the insurance part has an equity stake in that same firm? At this moment, the interests of the two parties are flatly opposed to each other.

To overcome these conflicts of interest *Chinese walls* will have to be built. It is our firm believe that more than any Chinese walls, a conduct of ethical behaviour will be necessary. Whatever Chinese walls may be established within an integrated firm, there will always be inside information, certainly at the top level. Only strict ethical behaviour with a firm top-down control will lead to a good risk management in this respect.

Internal conflicts of interest and sub-optimality

Not only the external effects, but also conflicts of interest may lead to a number of internal problems such as:

- struggle between the business units to get market share;
- danger of cannibalisation by simply switching portfolios (e.g. from the bank to the insurer or vice versa) with extra costs and without extra value added of profit or altering the growth perspectives of the existing (profitable) portfolio in one business unit by offensive moves for new (less profitable) market share by other units;
- cross-subsidisation and marginal tarification so that make-or-buy decisions are not in accordance with competitive market conditions or scarce resources are not always allocated in the most optimal, efficient or effective way:

  "the profitable businesses have to bear the financial burden imposed by other activities; if these attractive businesses were run alone, they could generate superior gains... the least profitable businesses are often overvalued because they only support a portion of their financial needs" (Very, 1993);
- struggle between different distribution outlets;
- cultural conflicts and image effects (difference towards risk aversion, different time horizons, selling techniques, remuneration structures, etc.

Quality downgrading

The overall quality level can be downgraded because some parts of the diversified scope of businesses are not well controlled; the service chain will be as strong as the weakest of its elements; the number of products a distributor or selling representative or agent can tackle successfully is not unlimited ! The limitation of the optimal scope of the product portfolio is not so much dictated by the consumers but by the operational limits and the limits of professional capacities. Therefore we do not agree with Robens (1987) that the crucial strategic factor is:

> "... die gesamte Bandbreite denkbarer Anlage-, Finanzierungs- und Versicherungswünsche, die von einem Individuen ausgehen (können)."

Abuse of power

Another interesting aspect is the argument that financial conglomerates would be(come) very powerful. Because concentration would be higher, financial conglomerates could abuse their power. On the contrary, some researchers in the US and in Canada argue that the whole development would enhance competition. By breaking down the barriers between the different financial pillars, a larger financial market is created whereby the number of suppliers will be increased to some extent.

THE OPINION OF THE SUPERVISORY AUTHORITIES: THE EUROPEAN COMMISSION[10]

The specialisation principle

The supervisory point of view on financial conglomerates implies an attitude of reserve and restraint. Because it is not the primary task of a supervisor to determine or to structure the market. The supervisory regulations have to follow the fundamental drifts of economic life.

The European Commission believes that financial conglomerates do present a number of problems from a prudential viewpoint. Some of these are linked to group structures in general, and some are specific to those heterogeneous groups, i.e. financial conglomerates (refered to as *those heterogeneous groups*).

Although it takes extremely varied forms, financial conglomeration is one of the striking features of today's financial markets. It corresponds to certain strategies and more importantly to certain needs: external growth is a never-ending trend in the face of

[10] This paragraph is based on a presentations by P. Clarotti, Head of unit Banks and Financial Institutions at the European Commission, and A.J. Vermaat, President of De Verzekeringskamer (NL), *in* Van den Berghe, L.A.A. (ed.), "From Bancassurance and Assurfinance towards All Finance", conference proceedings, March 19-20th, 1996, Brussels.

hyper-competition and decreasing profit margins. Therefore the Commission is certainly not against financial conglomerates, and wishes to interfere as little as is reasonably possible with this market-driven evolution. This statement was also confirmed by A.J. Vermaat, president of the Dutch insurance supervisor, who considered the trend towards financial conglomeration as some form of related diversification.

There are three main reasons for the rigorous *separation* of the credit, investment and insurance activities:

- the differences between the commitments respectively made to clients suppose different technical constraints and entail different financial and investment policies;
- insurance premiums should not be used to cover loans made by bankers to insolvent debtors, and bank deposits should not be used to finance the settlement of insurance losses (this, in the interests of the clients and of sound business practices);
- the third reason is that this separation would facilitate supervision by state authorities.

This specialisation principle is applied throughout the world, to various degrees. It exists in the U.S. and in Japan. It is also entrenched in EC Directives: the *Second Banking Directive* of December 1989 explicitly says that:

> "the Member States shall prohibit persons or undertakings that are not credit institutions from carrying on the business of taking deposits or other repayable funds from the public."

On the other hand, the *first Non-Life and Life Insurance Directives* (respectively 1973 and 1979 -as amended by the *third Directives* of 1992) say that:

> "Member States shall require every insurance undertaking for which authorisation is sought to [...] limit its business activities to the business of insurance and operations arising directly therefrom to the exclusion of all other commercial business."

More than that, life insurance undertakings must limit their business to life insurance (and ancillary non-life insurance business).

This specialisation principle was also found at the level of the supervisory authorities: we used to have a situation of segmented supervisors preponderantly managing on a solo basis.

Specific problems arise in group configurations

However, the constitution of groups raises specific concerns, as groups may blur the neat picture which would otherwise be derived from the specialisation of financial undertakings. On the other hand, groups also result in the leveraging of capital and, often, in an increase of certain exposures. These two different categories of problems have been treated differently at Community level.

Can banks own insurers and vice-virsa? There exist certain restrictions in the EC banking and insurance Directives. More precisely, the Second Banking Directive imposes limits to the participations of banks in non-bank institutions: those participations may not exceed 15 per cent of the bank's own funds in the case of a participation in an individual non-bank undertaking. The total of such participations may not exceed 60 per cent of the bank's own funds. However, Member States are totally free not to impose limits to participations in insurance undertakings. There is also no EC restriction on participations of insurance undertakings in banks.

It is therefore for Member States themselves to decide whether or not to impose limits to bank-insurance groupings. It is interesting to note that some used to restrict such groupings in the past, but that today most restrictions have been lifted.

The specialisation principle therefore does not apply through participations. This situation is well-anchored and the Commission has no intention of questioning that. This raises however a number of problems which will be discussed in the next points.

Applicable and proposed legislation for homogeneous groups: legislation for banks It were the bank supervisors who have started to respond to the challenges posed by conglomerates. They were primarily concerned with the likelihood of systemic risk (the risk that loss of confidence in one bank may be transmitted through the banking system, via the interbank market or the payment systems) and, as outlined above, contagion risk is, in reality, twofold:

- of a financial nature: because when a bank meets financial difficulties, an insufficiency of own funds can be transmitted to another bank which holds a participation in it, due to the depleted value of the participation held;
- of a psychological nature: because a loss of confidence in one company of a financial group might lead to loss of confidence in a parent or an affiliate.

The concerns of the bank supervisors' over the health of (international) banking groups was stimulated by the Herstatt failure of 1974, which led to the Basle Concordat, which *inter alia* spelt out the bank supervisors' preference for consolidated supervision of banking groups.

This technique deals directly with the contagion problem: by calculating a single own funds requirement for the whole group, it corresponds to replacing the book value of a bank's participations in other banks with the real prudential value of these participations, which takes into account the fact that these banks are themselves subject to own funds requirements. A single set of limits to risk concentration is also imposed. Finally, the systemic risk is indirectly tackled (although of course not entirely solved) since the need for externally generated own funds is higher for the consolidated group as a whole than for the sum of the single banks composing it.

Applicable and proposed legislation for homogeneous groups: legislation for insurance companies As regards groups of insurance undertakings a proposal for a Directive is on the table of the European Parliament and the Council. This new Directive will be discussed in more detail in chapter 10.

Supervising the insurance sector from a wider group perspective is not a new idea either: the collapse of Vehicle and General in 1974 is one famous example which took place more than 20 years ago. More recent examples are Hafnia and Baltica.

Possible approaches to financial conglomerates: national measures and international supervisory fora

Given the large increase in the number and size of bank-insurance type conglomerates in the EC, it is not surprising that the supervision of financial conglomerates is receiving widespread attention. There has been a number of international organisations that have been actively examining this issue since the early 1990s: in addition to the OECD and to IOSCO, the European Commission and the Basle Committee of Banking Supervisors are working on the prudential supervision of financial conglomerates.

Some Member States of the European Union have set up provisions aimed at dealing with this issue, for example Denmark in the form of capital requirements, Belgium and the Netherlands in the form of protocols between the different supervisory authorities involved.

What is the Commission's view on this issue?

Supervisory concerns and possible approaches

The prevention of double gearing This issue is one of the most fundamental quantitative issues raised in the presence of financial conglomerates. The problem of double gearing is not specific to financial conglomerates (see also the analysis of homogeneous financial groups). What is specific to financial conglomerates however is the complexity of the possible solutions. As regards the banking sector, consolidated supervision is the rule in the EC. As regards the insurance sector (see also chapter 10) the proposal for a Directive on Insurance Groups describes three different methods to eliminate double gearing, despite the homogeneity (by definition) of insurance groups. Only one of these methods starts from the consolidated accounts.

This gives a taste of the difficulty of finding both a coherent and a satisfactory way to eliminate double gearing in a group where a bank holds a participation in an insurance company or vice versa. The difficulty arises from the fact that banking and insurance rules regarding own funds play different roles within their respective fields, cover different risks and have been built up along different perspectives (see Van den Berghe, 1995). The absence of specialisation at group level creates this vexing difficulty.

The Commission already has some detailed indications on which methods could be practicable and how they could work. There are important reports (one from the *mixed technical expert group* set up by the Commission, and one from the *de Swaan group* set up by the Basle Committee). But the Community is far from having closed the examination concerning the elimination of double gearing in (heterogeneous) financial conglomerates.

Excessive risk concentration This is the second quantitative issue. All three financial sectors have rules both at national and EC level to limit excessive concentration of risk ("don't put all your eggs in one basket"). In other words in the banking sector we have the Large Exposures Directive which came into operation on 1st January 1994. This limits exposures to a single client or group of connected clients to 40 per cent of own funds, falling to 25 per cent of own funds after a transitional period. For investment firms we have somewhat more flexible rules in the CAD with higher capital penalties on exposures exceeding the 25 per cent figure after an initial 10 day window. In insurance we have the diversification rules relating to the investment of assets, representing the technical reserves.

The question which arises for a financial conglomerate is whether it matters that there are these conflicting rules and if so, how one should resolve the differences. For example: an insurance company cannot hold more than 5 per cent of its assets backing mathematical reserves in securities issued by a single company; but the so called *free assets* (more particularly those assets which are the counterpart of the solvency margin) might entirely consist of the said securities. Thus a bank could see its exposure to a single client significantly increased through its participation in an insurance undertaking. The Commission has not closed the debate on how best to tackle it. And again, the absence of specialisation at group level generates this difficulty.

Transparency of group structures Transparency of group structures is one of the core qualitative issues. Here the BCCI case is a prime example where through a holding company based in Luxembourg the bank operated through subsidiaries based in Luxembourg (and the Cayman Islands) but with the main banking activity taking place in branches in London and elsewhere.

The legislation of all three sectors has been amended by the post-BCCI Directive of June 1995, so as to formally empower bank, insurance and securities supervisors to revoke or refuse a license if they feel that a particular structure is no longer sufficiently transparent to allow effective supervision.

The problem is in fact not specific to financial conglomerates. It is not even specific to groups. But by requiring the presence in the same Member States of an undertaking's head office and registered office, this Directive has the effect that corporate structures will have to be clearer, as registered offices aren't allowed to be mere shells. In addition, refusal or even withdrawal of authorisation in the event that close links between a financial undertaking and another natural or legal person would prevent effective supervision has a concrete meaning for financial conglomerates.

Fitness and properness of shareholders In EC banking and insurance legislation, the requirement that major shareholders be *fit and proper* is clearly laid down. But what happens if the bank or insurance company is controlled by a holding company whose management doesn't have the required experience? Also, in a large conglomerate, the power of a single large shareholder can be greatly enhanced because of the dispersion of shareholders sometimes observed in groups.

Hence the need for greater vigilance to ensure compliance with the fit and proper criteria. The Commission is presently working on the question whether existing Community legislation is sufficient for the supervisory authorities to apply the fit and proper test in such a way that it is adapted to the sometimes very intricate structure of financial conglomerates.

Role and responsibility of the external auditors The concern here is twofold. First, if a number of auditors are reporting on different parts of the conglomerate, no one may have a single overview of the group's activities. Secondly to whom should auditors report any concerns they may have about the operation or viability of the conglomerate?

In response to the first point, it has been suggested to appoint a lead auditor who would be responsible for reporting on the whole group's activities. No EC legislation has been passed to that effect. As regards the second point (to strengthen reporting obligations) the post-BCCI Directive provides that external auditors should be obliged by law to mention any concerns they have to the appropriate supervisor.

Co-ordination between supervisory authorities The problem for the supervisors is how to secure an overall view of the risks run by a conglomerate. Each bank or insurance supervisor is responsible for applying capital requirements and other prudential controls in its own sector, but it will not necessarily know what is going on the other side of the fence. And this provokes the concern that risks run within a financial conglomerate may not be adequately addressed by any of the supervisors involved.

In the banking sector all legal obstacles preventing the exchange of information between supervisors have already been removed. Insurance supervisors have agreed to be bound by the same rules, and the corresponding arrangements are stipulated in the proposal for a Directive on Insurance Groups (see chapter 10). Promoting the full exchange of information between all supervisors will be essential for effective supervision. Regulators from one sector should be able to have a full picture of problems which may exist in other sectors. They need it for various reasons:

- to be able to assess the real value of a participation;
- to be in a position to foresee and thus prevent the negative impact of conflicts of interests within the financial conglomerate;
- to get relevant information as to the fitness and properness of management, etc.

The Commission is now working -and so is the *Joint Forum* (the new *de Swaan group*)- on the task to ensure co-ordination between supervisory authorities in a way that would be both efficient and very flexible. As regards the Commission, while doing this, two problems must be distinguished: co-operation between supervisory authorities in the same Member State, and co-operation between supervisory authorities in the different Member States. The Commission's role is of course more evident in the second case.

Monitoring of intra-group transactions Financial conglomerates are usually created to generate synergies between hitherto parallel lines of business. But the linkages are not always benign. Not only can combined operations create new prudential risks or exacerbate existing ones (which we have already seen), but also can these linkages lead to transfers of risks or of cash through intra-group transactions.

The Commission is presently working on the question of how, in the context of a financial conglomerate, they can allow the supervisory authorities to *monitor intra-group transactions* or in other words how they could be informed of these transactions and how they could have the possibility to restrict them. The proposal for a directive on insurance groups is showing the way to be followed (for more details see chapter 10).

Conclusion

These statements demonstrate that the European Commission believes that there are real supervisory problems in the growth of financial conglomerates. However financial conglomerates are there to stay: they represent a general trend for financial services in Europe. The supervisory authorities are answering to these new market trends with a certain time lag (see also point 10.2.1) and are well aware of the extreme difficulties of the exercise (this holds for the Commission as well as for the Joint Forum).

Supervisory responses must of course be adapted and need to be manifold. For example the Commission has to develop further the necessity to set up internal control procedures and reporting mechanisms. Another issue is the scope of the different measures: for example, the need to monitor intra-group transactions is more particularly associated with the existence of an influence exercised by one undertaking over another one (even if there is no capital participation), whereas the elimination of double gearing is more particularly associated with the existence of a capital participation (irrespective of the influence).

In any case, the Commission has no intention to hinder financial conglomerates, and considers that in general the supervisory rules should not force the undertakings to adapt the corporate structures that they have chosen unless this is essential for exercising effective supervision.

CONVERGENCE OF FINANCIAL INSTITUTIONS: THE U.S. ENVIRONMENT[11]

In the U.S. environment very little diversification exists as a result of banking and insurance affiliations. The fundamental reason for this situation is that the U.S. remains one of a handful of industrial nations with laws restricting banks from most involvement in the insurance business. Many believe these laws are archaic, contradictory and make little sense in today's environment.

Nevertheless there are significant attempts now underway to modify the U.S. regulatory environment that now requires the separation of banking and insurance operations and other ways are being used by financial organisations, particularly large commercial insurance companies and investment banks, to achieve more and more convergence in the U.S. marketplace.

U.S. regulatory environment

Legal barriers, separating banks from insurance are eroding significantly in the U.S. This is occurring at both the state and federal levels, and especially through court decisions interpreting the powers of national banks very broadly, as growing bank pressures gradually but steadily chip away at legal barriers.

---

[11] This paragraph is based on a presentation by E. Thrower, President of the College of Insurance New York, *in* Van den Berghe, L.A.A. (ed.), "From Bancassurance and Assurfinance towards All Finance", conference proceedings, March 19-20th, 1996, Brussels. The recent changes have been amended with the help of Michael Bradfield, partner of Jones, Day, Beavis & Pogue (U.S.).

However significant barriers still remain, particularly the Bank Holding Company Act ('BHC Act') which requires the separation of commercial firms from commercial banks. This federal law is widely regarded as the most significant law in the U.S. separating banks and insurance.

The BHC Act essentially limits banks from engaging in securities activities (together with the Glass-Steagall Act), and in insurance operations. The Glass-Steagall law was passed in 1933, during the depression era and following the banking crisis. An obvious question might be, why, after 60 years, is such a law still in place? The BHC Act was initially adopted in the 1950s and strengthened in the 1960s and again in 1970. There are two key reasons why these laws have not been repealed. First, the past attitude of many U.S. lawmakers has been supportive of strict federal regulations governing banking operations, a tradition that, in the past, has had strong public support. In addition, and much more important in the last fifteen years, bank entry into insurance has been fiercely opposed by insurance agents, especially independent agents, who fear their very survival if banks are allowed to enter insurance and agent groups have actively and successfully worked to keep restrictive laws on the books.

Most recently -in 1997- influence by agents, however, has diminished for several reasons. First, the attitudes of lawmakers are becoming much more supportive of greater deregulation of the financial sector as competition has increased and de facto combinations as well as regulatory loopholes have left the legal structure separating banking, securities and insurance in tatters. In addition, pressure is continuing to mount from banks, securities and commercial groups to repeal the Glass-Steagall Act and to substantially modify the BHC Act. Equally important, leading insurers have also begun to pull back from their historical support of laws restricting bank entry into insurance, and they are beginning to acquire thrift institutions that can take deposits just like banks.

These pressures are intensifying the battle between agents, banks and lawmakers regarding the future of the Glass-Steagall Act and the BHC Act. In 1997, the U.S. House Banking Committee passed a bill that would have allowed banks, investment banks and insurance companies to purchase each other with appropriate changes to Glass-Steagall and the BHC Act to allow this to happen. But this legislation, which had general Administration support, was not acted on by the full House in 1997 because of an inability to obtain a consensus among competing industry groups, or for that matter among the banking industry as a whole. While a general consensus exists on allowing combinations of banking, insurance and securities firms, wide divisions still remain especially over whether the reform should be more broadly based to remove all restrictions on the ability of commercial firms to own banks, and the outlook for comprehensive legislation remains very cloudy.

In the above discussion, reference was made to important court decisions which have helped to shape the debate in the United States about bank and insurance company affiliations. A major blow was sustained by those who oppose bank entry into the insurance business when the Supreme Court decided that the states cannot prevent national banks from selling insurance within their borders. The case at hand involved the state of Florida, one of twelve states barring bank operations in insurance, and a conflicting federal banking law that allows national banks to sell insurance in small towns with populations of 5,000 or less.

In this specific case, Barnett Banks, a large regional banking group, had been selling insurance in small Florida towns, based on federal banking law provisions. The state of Florida contended that state insurance laws restrict this activity and that such laws supersede federal banking laws. The Supreme Court decision nullifying the state law has had far-reaching implications for the bank-insurance affiliation debate and is a strong force for a fundamental change in the BHC Act. Another similar important step towards reform involves a 1995 Supreme Court decision that upheld the right of banks to sell annuities. This decision has opened the door for banks to gain a significant and increasing market share in annuity sales.

While it is clear that progress is being made, banks still have a long way to go in their movement into insurance. Their efforts are supported by a growing number of insurers, particularly those in the life insurance area who are initiating and developing partnerships with banks for insurance distribution. As long as the BHC Act remains unmodified, however, possibilities for any significant change are limited.

Convergence between the commercial insurers and investment banks

There is a growing convergence trend underway in U.S. financial markets involving commercial insurance companies, reinsurance companies and leading investment banks. Certainly, this trend is not unique to the U.S. It is, however, beginning to change significantly the very nature of risk management and the methods of financing risk. This trend is driven largely by corporate needs and demands for more effective types of financial protection for a broad range of financial and non-financial risks.

In the U.S., providing financing arrangements for corporate risks is becoming big business in the investment banking industry. Virtually all large investment banks in the U.S. have now developed large risk management divisions that focus on new products that utilise capital market tools for financing a broad array of corporate financial risks.

Investments banks, insurance companies, reinsurers and insurance brokers are also coming together to jointly form catastrophic property and liability insurance companies to write high limits of traditional insurance. Some examples include significant capitalisations from J.P. Morgan for the formation of Bermuda based EXEL, Limited and Mid-Ocean Reinsurance, and Morgan Stanley's investment in ACE, Limited.

Travelers Group is perhaps the most broadly integrated U.S.

insurer in this regard, with ownership of a major investment banking firm, Smith Barney Holdings. In addition to providing investment opportunities, Travelers now offers diversified financial services that include life and annuity products (through ownership of Primerica), commercial financing (through ownership of Commercial Credit), property and casualty commercial insurance (through ownership of Travelers Insurance) and now, personal lines property and casualty insurance (through the new $1.0 billion purchase of Aetna Property & Casualty). Travelers is further developing highly integrated cross-selling activities involving Smith Barney brokers, Commercial Credit representatives, and the Primerica Life Insurance sales force, which comprises over 100,000 agents, most of whom are part-time. It appears that Travelers now has most of the pieces in place for a broad based financial services operation except for ownership of a commercial bank. As noted above, these insurance-financial conglomerates are now moving into the depository business through the acquisition of thrift institutions in the U.S.

Large insurers have also been expanding into this area. Recently American International Group established AIG Capital Management, a division focusing on global investment banking. In 1992, Marsh & McLennan established a new division, Marsh & McLennan Risk Capital, to focus specifically on investment activities, primarily in the insurance industry. Marsh & McLennan has also gained an increasing foothold in insurance underwriting, with significant investments in companies such as ACE Limited, EXEL, Centre Reinsurance Holding Limited and Mid Ocean, Ltd.

One example of the expansion of investment bankers into risk financing is the contract recently awarded to J.P. Morgan for the financing of Nationwide's catastrophic risk exposures. This took the form of a $400 million surplus note transaction, and is an alternative to traditional reinsurance products.

One final example involves Chase Global Financing Division, with 50 professionals devoted solely to securing corporate risk financial needs. In five years this division has reported completing $25 million in insurance related financing transactions. Recently they established a $500 million credit facility as part of a $1.6 billion Hawaiian hurricane catastrophe fund. This facility is also an alternative to traditional reinsurance while offering broader financing at a lower cost.

In conclusion, many new developments are underway in the convergence of commercial insurance, reinsurance and investment banking. While it remains uncertain just how extensive this trend will become, probably we will see more and more innovative blending of the products, services and operations of the two sectors. This convergence provides compelling evidence that the U.S. commercial insurance market will not remain the sole province of traditional insurers and reinsurers.

It will be interesting watching how these sectors converge in the future and the impact this will have on the very environment of insurance and risk management.

10

TOWARDS A PROPOSAL FOR A EUROPEAN DIRECTIVE ON FINANCIAL CONGLOMERATES

HISTORICAL BACKGROUND

The European Commission as well as the supervisory authorities of the banking and insurance sectors are concerned about the stability and solvency of *diversified* groups within the financial sector. The European commission as well as the Economic and Social Committee state that the increasingly close ties between insurance companies and other financial institutions within the *so-called* financial conglomerates mean that appropriate prudential measures are necessary in this respect. At the same time they are looking for a *level playing field* which also encompasses comparable or even integrated approaches towards solvency regulation and supervisory mechanisms.

As a result of the fact that the banks are supervised on a consolidated base whereas the insurance sector opted for a separate control per legal entity (whether an independent company or a partner within a larger group or conglomerate), it is more difficult to approach the supervision of financial conglomerates.

The question whether the supervision of the insurance sector should be focused on the group level or rather on the level of the individual companies has been in open debate for a number of years. The growth of the number of financial conglomerates forced the insurance sector to come to a consensus on the way towards an extra level of supervision on top of the solo-supervision which is the so-called solo-plus approach[12].

Although the ultimate goal is to establish a co-ordination of the supervisory instruments and the regulatory systems, the first step will be to amend the regulation of insurance companies to complement the actual regulation of insurance companies with a

---

[12] These concepts have been explained in detail in Van den Berghe, L.A.A. (ed)., "Financial Conglomerates: New Rules for New Players?", Kluwer Academic Publishers, Dordrecht, 1995.

group component. The first draft proposal dates back from 1995 (Com (95) 0406-C4-0464/95). In order to be able to analyse this proposal, the following questions have to be examined:

- Which companies and groups will fall under the application of these Directives. This means: what type of activities are concerned and when is a company part of the supervision on the solo-plus group level?
- What are the areas of supplementary supervision and what extra rules are introduced?

Although it is premature to make a final evaluation it is certainly interesting to look at the potential direction of this extra regulation[13].

PROPOSAL FOR A SUPPLEMENTARY SUPERVISION ON INSURANCE GROUPS BY THE EUROPEAN COMMISSION

Scope of the proposal

The present debate is confined to *insurance groups proper*, i.e. groups which are composed primarily of insurance undertakings. Article 1 makes a distinction between an *insurance holding company*, whose subsidiaries are exclusively or mainly insurance or reinsurance undertakings, and a *mixed activity insurance holding company*, whose subsidiaries include at least one insurance undertaking but whose parent company is neither an insurance holding company nor an insurance undertaking. The first category is subject to all extra regulatory measures whereas article 3 explicitly states that *mixed activity insurance holding companies* are subject to certain aspects of the supplementary supervision[14].

The undertakings falling under the solo-plus supervision rules only are insurance undertakings having a registered office in the Community. Unlike the banking sector, where many standards are elaborated at global level (in fora such as the World Bank, the IMF, BIS, etc.) the insurance sector has been lacking such a world-wide approach for quite a long time.

The *tripartite* approach to financial conglomerates shows that a more global approach -in addition to the Community approach- may also pay dividends for the insurance sector.

---

13 This analysis is based on detailed research in order to prepare an Expert's Report for the European Parliament.

14 Mixed activity insurance holding companies are not subject to the double gearing provisions but they have to obey to the principles of information and supervision of intra-group relations.

When is a company subjected to the solo-plus supervision and how are insurance groups defined?

According to the proposal for an EU-Directive there is a three-pronged approach.

(1) A definition based on the 7th Company Law Directive (83/349/EEC). This concerns insurance groups where there is a *parent-subsidiary relationship*. Various types of relationships can be identified in this connection:

(a) the parent undertaking has the majority of the voting rights in the subsidiary;

(b) the parent (as shareholder/member) can appoint the majority of the executive directors etc. of the subsidiary;

(c) there is a dominant influence (of the parent-shareholder/member in the subsidiary);

(d) a majority interest (in accordance with points (a) or (b)) can also be acquired by means of co-operation agreements (shareholder agreements); in this case, a minimum participation of 20% (of the voting rights) may serve as a condition.

In case of points (a), (b) and (d) there is a quantifiable criterion. Point (c), however, requires an assessment of an actual situation. For the purpose of determining the group relationships, both direct and indirect participations may be taken into account (e.g. a subsidiary of a subsidiary is also regarded as a subsidiary of the parent company).

(2) In addition to the definition of parent-subsidiary relationships under the 7th Directive, the Proposal for a Directive also provides the possibility that in case of insurance groups the supervisory authorities should decide for themselves when the actual situation with regard to control amounts to a *dominant influence*. This extra provision seems unnecessary since point (1.c) already covers this.

(3) A definition based on the 4th Company Law Directive (78/660/EEC) and the Directive on Credit Institutions (92/30/EEC) provides that relationships between insurance and reinsurance undertakings and holding companies which may be described as a *participating interest* are also considered to be insurance groups. These are participations -other than those, referred to at point (1) and (2)- which fulfil the following criteria:

(a) a significant influence may be exerted on the activities of the associated undertaking;

(b) there is a durable link, which is in our view rather different from a significant link, as referred to in the present Proposal for a Directive;

(c) the aim of this link is to stimulate and to contribute to the undertaking's own business operations.

The 4th Directive gives the member states the freedom to determine the threshold from the moment at which there is a *presumption* of a *participating* interest. Under the 4th Directive, however, this limit may not be higher than 20%.

The Directive on the consolidation of credit institutions (92/30/EEC -article 1) defines *interest* effectively as the (direct or indirect) ownership of 20% or more of the voting rights or the capital of an undertaking.

This provision has been extended to insurance undertakings in the Proposal for a Directive on insurance groups (article 1). It is, however, proposed to make a temporary derogation: from the introduction of the Directive (which was originally scheduled for July 1st, 1997) until July 1st, 2001, the limit should be set at 25%[15].

Critical evaluation of proposals Although this approach runs entirely parallel with the corresponding provisions for credit institutions (see point above), the difference between banks and insurance companies may not be disregarded.

As institutional investors, insurers are obliged to invest their resources as efficiently as possible. Indeed, there are detailed rules governing the investment of the technical reserves. The equity capital must be invested in the best possible way, both for reasons of solvency and competitiveness. It follows that insurers -particularly the diversified and large or relatively large insurance groups- have, directly and indirectly, many investments which they do not regard as a participation and which cannot be classified as a participation on the basis of the criteria laid down in the 4th Company Law Directive (78/660/EEC).

Whereas explicit reference is made in the definition of parent/subsidiary relationships to evaluation by the supervisory authorities of the *actual* situation, this evaluation is completely missing in the definition of *participating interest.* This seems to be a serious defect in the present Proposal for the Directive.

Another point we do not really agree with is the way in which provisions from general directives (the 4th and 7th Company Law Directives -78/660/EEC and 83/349/EEC)- and the Directive on credit institutions (92/30/EEC) have simply been transposed to the insurance sector. (Since insurers are by definition institutional investors, the definitions of *group, parent/subsidiary undertaking* and *participation* certainly require special attention). In our view, it is not appropriate simply to apply the general definitions to

---

15 If the recommendations of the Economic and Social Committee were to be followed, this temporary derogation would no longer be necessary (opinion, point 3.3.1.6.2).

insurance undertakings.

(There are also considerable differences between the banking and insurance sectors, particularly for what concerns the *money-creating* potential and the *diverse* categories of security capital).

In brief, the main differences are as follows :

- Money-creating function:

  In the financial and monetary economy, the financial institutions are classified as having the capacity of money creation. Monetary institutions can have a money-creating function. Their liabilities, for example bank accounts, represent purchasing power. It is on the basis of these liabilities that they can in turn lend money, and such loans once again represent purchasing power. By contrast, non-monetary institutions cannot engage in money creation. This is why monetary institutions can also have a much greater impact on the financial stability of a country and why they are exposed to a much greater danger not only as regards solvency but also as regards liquidity. This also explains why much greater importance is attached to the banking institutions in the control of monetary policy. Whereas banks qualify as monetary institutions, insurance undertakings are classified as non-monetary institutions.

- Differences in terms of capital base:

  Banks have only their equity capital as a solvency buffer; insurers, on the other hand, have two forms of capital base: besides their equity capital they have extra buffers in their technical provisions.

What are the areas of supplementary supervision?

The proposal for a directive contains three groups of additional measures.

Improvement of the exchange of information In order to be able to exercise supervision over insurance groups in an effective way, it is necessary that the supervisory authority should have access to all relevant information. For this purpose the supervisor has a right of inspection (or droit de suite) not only in all undertakings with which the insurance company has a parent-subsidiary relationship but also in every undertaking in which the insurance company has a participating interest[16].

The relevant information is defined as follows:

---

16 The droit de suite does therefore not apply to *upstream* relationships, in particular undertakings which have a holding in insurance undertakings.

Article 5:

guaranteeing the availability and quality of additional information by developing or having the disposal of appropriate internal supervision mechanisms (paragraph 1) and removing any legal obstructions to the mutual exchange of information (paragraph 2).

Article 6:

access to information for competent authorities (*droit de suite*).

These measures apply to insurance and/or reinsurance groups, insurance holding companies as well as to mixed activity insurance holding companies.

Rules for transactions within a group The diversification requirements with regard to the assets which apply as cover for the technical provisions of insurance undertakings (3rd EEC Insurance Directives) have not been extended to the other assets because this is considered to be a restrictive solution which can have undesired side-effects. As an alternative, it is proposed that there should be supervision of significant intra-group transactions, which must be effected at arm's length and must be reported retrospectively to the competent authorities on an annual basis (art. 8). These measures apply again to both specialised as mixed activity groups.

Various objections can be formulated against this proposal:

- The description of the intended transactions seems very complex. The Economic and Social Committee has commented on this (see point 4.6). Clarification seems necessary, not only with regard to the *exhaustive list* -loans, guarantees of off-balance-sheet items, investments, capital assets for the solvency margin and reinsurance- but also concerning the following matters.
- The use of the terms *market conditions* and at *arm's length* may raise questions. The Economic and Social Committee has also commented on this; it proposes that the expression *according to prevailing usage* be employed. However, the comment on reinsurance transactions (4.6.1) does give rise to questions. For example, is it indeed correct that different principles about transfer prices should be applied in the insurance industry? On what grounds should this be done? And to what extent do combinations of *insurance and reinsurance undertakings* differ from other forms of vertical integration within groups of undertakings?
- Notwithstanding the fact that only *significant* transactions need to be reported, this proposal appears to raise questions. What does *significant* mean? For whom must this be significant? Here the suggestion of the Economic and Social Committee seems very worthwhile (4.6.2.1).

- As the proposal now reads, there is a *droit de suite* in case of both the undertaking which has participating interests and the undertakings in which such participations are held. This seems a rather complex arrangement. Could it not be solved by providing that additional supervision may be exercised only with respect to the participating undertaking? If there are cross-participations, each partner will have an interest in the other, and additional supervision of both will therefore be necessary.

Is there not a danger that these new arrangements will entail fairly far-reaching intervention by the supervisory authorities? To what extent is this necessary in connection with a supplementary solvency supervision? And how does this regulation affect the pursuit a more free functioning of the market?

Rules to prevent double gearing In order to prevent double gearing article 9 defines adjusted solvency criterion whereas article 10 adapts this measure to the insurance holding companies. These extra regulations DO NOT apply however to mixed activity holdings. The complexity of this subject and the importance attached to it by the market players as well as the supervisory authorities justify a more in-depth analysis of these new rules.

HOW CAN DOUBLE GEARING BE AVOIDED?

Attempt at defining double gearing in more detail

The Proposal for a Directive defines double gearing in the following passage (art. 9):

> "The problem of double gearing can arise when an insurance undertaking acquires participations in undertakings. Double gearing is caused by using the same capital more than once to cover the capital requirement for insurance undertakings in the same group."

This definition seems both too narrow and too wide:

- too narrow because it merely refers to the double gearing between insurance undertakings;
- too wide because it makes no distinction between the creation of *notional* capital and the normal use of capital from an *investment perspective* or from the point of view of asset liability management.

Both propositions require further explanation.

Double gearing sensu stricto In our view, cross-participations give rise to the most extreme form of double gearing, where there is undeniably notional capital.

This can be clarified by an example:

- Situation 1:

  The equity capital of insurance company 1 is 100 million ECU and of insurance company 2 is 150 million ECU.

- Situation 2:

  For various reasons, the two insurers decide to arrange a cross-participation; for this purpose they each increase their capital by 25 million ECU, thereby raising the equity capital of insurer 1 to 125 million ECU and of insurer 2 to 175 million ECU.

Although application of the present solo-supervision criteria would suggest that there has been an (apparent) improvement in the solvency position of both companies, the solvency is not intrinsically higher. Clearly, this form of double gearing must result in corrections of the solvency margin. This does, however, not mean that double gearing is totally undesirable for reasons of business economics (Van den Berghe, 1995).

- From the point of view of capital efficiency and in order to achieve the maximum leverage from the capital position, undertakings may look for methods to use capital in more than one way within a group structure; cross-participations and other capital relationships between and with subsidiaries and sister companies can mean that the *net* or *consolidated* solvency position is lower than the gross aggregated balance sheet items would suggest.
- To eliminate this *notional* capital, the Proposal for a Directive therefore rightly suggests the adoption of an *adjusted* solvency position of such kind that double gearing (or rather this form of double gearing) can no longer occur.

Double gearing sensu lato Double gearing often occurs in a modern, financially-developed society, without necessarily involving notional capital. This is an acknowledged fact in the theory of financial intermediation: financial intermediaries create financial assets, engage in money creation etc. Therefore it is necessary to determine to what extent there is double gearing in the wider sense, or to what extent it merely concerns the phenomenon of financial intermediation.

Insurers are obliged -by virtue of their primary function of covering risk and because their operating cycle is inverted- to set aside technical reserves or provisions, which they must invest as efficient as possible. In addition to this insurance related *technical* capital, the solvency capital should also be used efficiently. This creates the well-known function of the institutional investors. In fact, this in itself entails the double use of capital. As a financial intermediary, they will recycle their equity capital -like their *technical* insurance related capital- in all kinds of investments. Whether this happens within a group or between unrelated businesses makes no immediate difference, at least theoretically. The most relevant question is to what extent investments or usage of funds can result in an increase of risk and therefore pose a threat to the solvency. This is why it is necessary to examine to what extent investment risks exist (such risks are naturally smaller in case of government securities than in case of venture capital) and to what extent there is a substantial correlation between the risks of the relevant insurance undertaking and the undertaking in which it invests. Is it better to invest in a brewery or a textile company than in another insurance undertaking? (Hesberg and Karten, 1994).

This is why we suggested that the definition of double gearing in the Proposal for the Directive is in some respects too narrow. As it stands, the definition merely covers the double use of capital to cover the capital requirements of insurance undertakings. Although this definition matches the field covered by the proposal, it should be pointed out that not all forms of double gearing are excluded by this. For example, what about entire or partial cross-participations with/of other undertakings? If these other undertakings have minimum capital requirements, this can result in notional capital accumulation. Most attention should be paid to financial institutions such as banks, securities institutions, etc.

It is, however, necessary to warn once again against extrapolating the provisions relating to credit institutions and extending them to insurance undertakings.

Since insurers are *institutional investors* their use of resources is of a totally different nature. In addition, detailed rules already exist with respect to the investments of their technical reserves, both regarding the composition as the valuation of these assets[17].

Would a modification of the solvency criteria in the direction of risk-based capital for all insurance undertakings under supervision not provide a way of combating risk escalation?

If, on the other hand, it is decided to *eliminate* double gearing in the insurance sector, it would seem to be necessary to review the definition of *participation* as suggested above. In other words, double gearing sensu lato should refer to the double use of capital within insurance groups, other than through cross-participations (double gearing sensu stricto), with the exclusion of uses of capital that may be classified as *investments*.

---

17 To what extent the remark of the Economic and Social Committee (point 2.3) that the problem of double gearing in the case of insurance groups is clearly different from that in the case of banks is also based on this proposition is not evident from its opinion.

What are the possible consequences of the different definitions of a group in practice?

The question whether this Directive could have an important impact on insurance markets is quite difficult to answer. At least to have some insight into the possible consequences -particularly for the implications of the definition of *participation* and *insurance groups*- some analyses were made. Although there exist numerous possibilities to determine to which groups separate companies belong, we have chosen a definition which suits the proposal of Directive. Therefore:

- we will not only take into account direct relations but indirect relations will be considered as well;
- we have used different limits when defining *participation* in order to be able to analyse the impact of these alternatives. We have chosen the following limits: 100%, larger than 50%, larger or equal than 33%, 25% and 20%;
- we have taken into account both offices within as outside the EU.

Tables 11.1 to 11.6 give a detailed picture of the group relations we found in our database of financial institutions. Table 11.1 to table 11.5 consider only group participations between firms located within the EU; in table 11.6 we have taken into account group relations with firms outside the EU.

It is clear that the number of group relations to be considered is increasing as the limit value of the relevant participation gets lower. A lower border doesn't only mean that the number of group relations increases but also that members of a group can act as a parent company themselves. This has everything to do with the fact that relations are assigned to a parent company in function of the most important shareholder. This company in turn can be a member of another group. If there are no assignments any more, we keep three kinds of companies: parent companies, group member and unrelated companies. If the new parent company has a new core activity then the group participation also changes.

If we take 20% as the limit percentage then in our database[18] we identified 1571 group relations within the EU. 478 can be considered as insurance groups, 41 are reinsurance groups and 487 are mixed banking and insurance groups. If we increase the limit to 25% or 33% then the number of relations decreases to 1532 resp. 1475. Less than 900 of these relations are full subsidiaries, i.e. there exist a 100% mother-daughter relationship. From these figures we can conclude that this certainly is not a *marginal* phenomenon. Even more, the assignment of group relations to

[18] This database does not cover the whole financial market. However for the Belgian and the Dutch market this database is quite accurate; for the other financial markets it gives a good picture of the most important players.

Figure 10.1

Within EU	P = 100%	Activity group member						
		Credit	Insurance	Reinsurance	Investment	Mixed	Other	Total
Activity group member	Credit	94	32	0	2	90	8	226
	Insurance	33	233	1	1	104	10	382
	Reinsurance	1	3	1	0	6	0	11
	Investment	14	21	0	2	38	0	75
	Mixed	0	2	0	0	4	2	8
	Other	31	46	0	3	109	3	192
	Total	173	337	2	8	351	23	**894**

Figure 10.2

Within EU	P > 50%	Activity group member						
		Credit	Insurance	Reinsurance	Investment	Mixed	Other	Total
Activity group member	Credit	137	54	0	4	94	38	327
	Insurance	68	291	6	2	105	59	531
	Reinsurance	1	4	8	1	5	2	21
	Investment	47	28	0	4	42	9	130
	Mixed	1	4	0	0	7	4	16
	Other	44	68	0	7	135	28	282
	Total	298	449	14	18	388	140	**1307**

Figure 10.3

Within EU	P ≥ 33%	Credit	Insurance	Reinsurance	Investment	Mixed	Other	Total
		Activity group member						
Activity group member	Credit	147	55	1	4	104	58	369
	Insurance	89	310	8	1	107	73	588
	Reinsurance	1	3	9	1	5	3	22
	Investment	41	29	0	1	52	22	145
	Mixed	1	4	0	0	7	4	16
	Other	56	64	0	17	161	37	335
	Total	335	465	18	24	436	197	**1475**

Figure 10.4

Within EU	P ≥ 25%	Credit	Insurance	Reinsurance	Investment	Mixed	Other	Total
		Activity group member						
Activity group member	Credit	130	73	1	4	113	63	384
	Insurance	88	295	22	1	106	88	600
	Reinsurance	1	3	10	1	5	3	23
	Investment	42	31	0	1	53	23	150
	Mixed	1	4	0	0	7	5	17
	Other	58	68	1	5	186	40	358
	Total	320	474	34	12	470	222	**1532**

Figure 10.5

Within EU	P ≥ 20%	Activity group member: Credit	Insurance	Reinsurance	Investment	Mixed	Other	Total
Activity group member	Credit	129	74	6	5	113	67	394
	Insurance	88	295	22	1	109	90	605
	Reinsurance	1	3	10	1	5	3	23
	Investment	43	33	0	1	55	23	155
	Mixed	1	4	0	0	7	5	17
	Other	61	69	3	5	198	41	377
	Total	323	478	41	13	487	229	**1571**

Figure 10.6

Within & Outside EU	P ≥ 20%	Activity group member: Credit	Insurance	Reinsurance	Investment	Mixed	Other	Total
Activity group member	Credit	202	65	0	5	116	58	446
	Insurance	76	368	8	2	154	73	681
	Reinsurance	1	7	13	1	7	2	31
	Investment	55	31	0	5	56	11	158
	Mixed	1	5	0	0	7	5	18
	Other	71	98	0	7	178	38	392
	Total	406	574	21	20	519	186	**1726**

groups indicates that there are more group relations and legal groups in the insurance sector than in the banking industry. This can partly be explained by the obligatory specialisation in a lot of countries.

These group relations become really complex in case of financial conglomerates: on average each insurance group consists of 6 companies; for the bank groups this figure is 4.5, but for financial conglomerates this figure is 81. Of course these figures should be interpreted with care because there is a bias towards these financial conglomerates. The database is focused on these financial conglomerates which may explain why we found these lower figures for pure banking or insurance groups.

Is the European insurance industry confronted with double gearing: a test

Although it is not possible to analyse this question in more detail it seems interesting to dwell on a little longer on the question of *double gearing*. On the base of the aforementioned database we have tried to investigate this question. We defined cross-participations as:

> "a collection of one or more direct participations which has the consequence that each of the companies involved (indirectly) is shareholder of itself."

In first instance, we only looked at cross-participations whereby all participations were known[19].

The total number of companies with cross-participations and thus double gearing s.s. is 105. The accompanying percentages vary quite strongly. It is interesting to point to the fact that this phenomenon is more active in some EU markets such as France and Belgium. However as this issue is quite complex it doesn't make much sense to analyse it on a macro-economic level but it is better to look at it for each group separately.

Neutralising the effect of double gearing and calculation of the adjusted *solvency margin*

The general principle is that when the required solvency margin of an insurance undertaking is calculated, account should be taken of the aggregate of the capital requirements of all related[20] insurance

---

[19] This means that for example a company with a cross-participation via two other companies is only included in the analysis if for all three companies the exact share is known.

[20] This includes associated or participating insurance or reinsurance undertakings both within the Community and in third countries (art. 9.2). The Economic and Social Committee does not think fit to impose these obligations on participating undertakings from *outside the EEA* (in the hope

undertakings on the one hand and of the *net* supervision capital available on the other. In determining this last component, double gearing must be excluded.

Insurance undertakings (which themselves possess participations in one or more insurance undertakings)
Three methods[21] (presently in use in a number of member states) are considered to be equivalent (art. 9 and Annex I to the Proposal[22]). In the absence of consensus between the different supervision systems it is possible that no European Directive will emerge. If the methods in question are indeed equivalent, the choice could in our view be left to the undertaking itself. It would then be possible for all parts of a particular group to apply the same method. This would not only help to cut costs and thus benefit consumers but also would allow a free choice to be made in open competition, without this constituting regulatory arbitrage.

Below are a few more brief observations about the three methods of calculation:

(a) Deduction and aggregation method (method 1)[23]

This is in our view the most laborious method of calculation. Both the solvency criteria and the determination of the available solvency capital are corrected to take account into the group relationships (this method seems to lead to very complicated calculations, especially in case of complex group relationships). The solvency situation of the *active* or participating undertaking is determined as the aggregate of the traditional assets (as determined in the solo supervision) and the proportionate part of the corresponding capital of the *passive* or related undertaking.

The solvency criteria of the *active* undertaking consist of 3 components:

- the classic solvency criteria of the *active* or participating undertaking (as calculated in case of solo supervision);
- the proportionate part in the equivalent solvency criteria of the *passive* or related undertaking;
- the book value in the active undertaking of all assets for the purpose of the solvency margin of the related or

that these proposals could be effectively applied to all EEA countries and that adequate co-operation could be achieved with the other OECD countries).

21 It is unclear which member states apply which methods. The provisions in the Opinion of the Economic and Social Committee (point 2.4) clearly differ from the survey results of the C.E.A. (19/9/96) - MU 024(02/96).

22 The structure and numbering of Annex I (Dutch version) deserves extra attention as the numbering does not seem logical.

23 The detailed provisions of Annex I seemed unclearly formulated; further explanation is certainly warranted.

passive undertaking.

Is it correct to state that as a result of this a surplus in the solvency position of the *passive* undertaking can indirectly benefit the adjusted solvency margin of the *active* undertaking?

(b) Requirement deduction method (method 2)

In case of the *active* or participating undertaking, only the *solvency criteria* are modified. The traditionally defined criteria (in accordance with the solo supervision) are supplemented by the *proportionate part* of the solvency criterion of the passive or related undertaking.

Not only is this method of calculation simpler but it also seems more neutral in terms of the influence of the solvency position of the related or passive undertaking on the solvency position of the active or participating undertaking.

(c) Consolidated annual accounts (method 3)

In this case the danger of double gearing is obviated; the required solvency margin is calculated with respect to the consolidated items. In so far as there is no solo supervision, the consolidated approach allows deficits to exist in some parts of the group which can then be set off against surpluses in other parts of the group; this shows that:

- the three methods are certainly not equivalent;
- solo supervision is certainly of value (see also later), particularly since the legal entity -the insurance undertaking under supervision- is the opposite contractual party of the policyholder and the latter has no recourse against the insurance group.

In our opinion, insufficient reasons have been given for altering the definition of the relevant assets in relation to the issued and unissued share capital and the profit reserves and the future profits of life assurance undertakings. We can therefore understand why the Economic and Social Committee has suggested that this limit should be dropped. We have the impression with respect to both this point and the following point that the Proposal for a Directive wishes to use the regulation of insurance groups to alter certain parts of the supervision of insurance undertakings. Since the proposed criteria apply only to insurance groups, there is a danger that competition may thereby be distorted to the detriment of insurance groups.

Reinsurance undertakings[24] The Proposal involves a

---

[24] These provisions do not detract from the fact that reinsurers are not subject to supervision imposed by the Community. Even where measures are necessary, it is proposed that they be aimed at the

tightening up of the solvency requirements for the member states which do not yet exercise supervision over reinsurance undertakings. Perhaps this explains the suggestion of the Economic and Social Committee that regulation of reinsurance undertakings should be left to the Member States.

Reinsurance undertakings that form part of a reinsurance group will have to comply with a *notional solvency requirement* through the direct insurers to which they are related. Since this can effectively constitute double gearing, in addition to the solvency criterion this would at first sight seem justified. However, one may wonder whether the problem of double gearing only appears when the reinsurer is a related undertaking. Can the same problem not occur if the reinsurer is a *participating* insurer?

Nonetheless, the insurance industry's criticism of this part of the proposal for a Directive seems to be well-founded: if supervision of reinsurance undertakings, that form part of a group, is tightened up, there is a clear danger that this may distort the competition with other (independent) reinsurers and particularly with reinsurers outside the Community. In view of the very international nature of the competition and the attraction of various tax havens, this potential threat seems very real. This is why regulation of the solvency of reinsurers should perhaps be arranged only on an international basis.

Insurance holding companies Two approaches are provided. Annex 1 of the proposed Directive refers to an intermediate insurance holding company, i.e. an insurance undertaking in which another insurance undertaking possesses an interest but which itself has an interest in another insurance or reinsurance undertaking. The three methods described above (point 1) remain applicable to the participating insurance undertaking. However, the question arises as to what extent the complex structure of some insurance and reinsurance groups permits an approach of this kind. Would it not be more effective, and certainly much simpler, if extensive supplementary supervision and adjusted solvency which takes into account all relevant group relationships were to be required only at the level of the top holding companies?

Annex 2 (and article 10) of the proposed Directive concern the insurance *top* holding company where the holding company is the parent company.[25] Here the holding is subject to a *capital test* which can be determined by 2 methods[26]:

- solvency warning test;
- test on the basis of consolidated annual accounts.

---

subsidiaries and not at the relevant reinsurer(s).

25 The description of this kind of holding is anything but clear in the opening words of Annex II.

26 The C.E.A. proposes that article 10 and Annex 10 be dropped and that a top insurance holding company should therefore not be subject to any capital test (MU 022/02/96).

In both cases the capital of the insurance holding company should equal or exceed the sum of the binding and non-binding solvency criteria of the related insurance undertakings or related reinsurance undertakings. According to the Opinion of the Economic and Social Committee, these criteria are less strict than the methods for the prevention of double gearing in case of the equity capital of insurance undertakings and carry lighter sanctions.

Related insurance undertakings in third countries If there are legal impediments to the transfer of the required information, the solvency situation of insurers established in the Community is reduced by the amount of the participating interest in the non-Community participation (and all other capital components eligible for the solvency margin of the related undertaking).

From solo to solo-plus supervision[27]

From an academic point of view, we wonder whether in a period of global competition and growing interconnection between companies, for example within the financial and insurance sector, it would not be advisable to switch to an evaluation on a more *consolidated* basis also for insurance groups. However this does not mean that different solvency criteria should not continue to be employed for life and non-life insurance companies. These criteria should in turn differ from the criteria in the banking sector. As already observed in relation to the supplementary solvency margins, solo-supervision is in this respect certainly most effective.

It would seem advisable to carry out a thorough investigation of the comparative economic costs and benefits of the present approach and an approach on a more consolidated basis. We are convinced -after studying thousands of relationships within the financial sector- that the interconnections are becoming so important that a *fragmented* (solo) approach will have to incorporate an increasing amount of extra supervision (supervision plus) if sufficient information is to be obtained about the complex reality of these relationships.

Studies have shown that a higher degree of integration is required in order to achieve latent synergies (on both the output and the input sides) in practice. The greater the integration, the greater will be the number of intra-group relationships. And the harder it will become to allocate the vague overhead structures to the various operating companies. This does not only apply to the so-called general overheads (e.g. overheads incurred in connection with cross-selling) but also to the allocation of investments (a field which is ideally suited for achieving scale and diversification benefits) and the corresponding investment income.

---

[27] The term used to cover this supplementary supervision is *solo-plus supervision*. However, the Opinion of the Economic & Social Committee refers to *solo supervision plus*.

One can also think of the growing integration of products (both life and non-life insurance) in general customer packages (e.g. employee benefits packages, general family policies, etc.). Market reality is drifting further and further away from the traditional legal division into separate financial sectors and their respective supervisory authorities. This is bound to result in an ever-widening gap between prudential supervision and the reality of what is happening in the market and the commercial operations.

According to this philosophy, it therefore could be permissible for *surpluses* to the required solvency margin within a group to be taken into account in the group undertakings, or that in other words some kind of system of *communicating vessels* would be established to respond flexibly to the dynamic expansion of certain parts of the group. This would mean, however, that a *recognised* standard could be applied by all undertakings in the Community.

SECTION III

SYNTHESIS AND CONCLUSIONS

11

SYNTHESIS AND CONCLUSIONS

1. New buzz words like bancassurance and all finance need to be clearly defined in order to differentiate between distribution issues and more structural diversification. We therefore made a clear distinction between bancassurance and assurfinance on the one hand and all finance and *financial conglomerates* on the other hand.

Banks were predominantly leading the diversification trend. Their diversification strategies were induced by structural changes at the level of the wholesale market (e.g. securitisation, desintermediation) as well as at the level of the retail market (changing patterns of savings and investments). Part of this diversification trend with banks has therefore originated in the insurance sector, which attracted an increasing proportion of the saving money (mainly in the form of pension-related savings).

At first insurers were collaborating to supply these banks with the appropriate insurance products thereby benefiting from an extra distribution channel. Later on also insurers have diversified into the direction of assurfinance or even opted for a full-fledged financial conglomerate. Many methods are used to realise the diversification strategy: starting from scratch, mergers and acquisitions, joint ventures and marketing agreements.

2. The first moves in the direction of bancassurance were mainly oriented towards the life insurance market, hereby recovering part of the lost ground in the savings market. But as success breeds success, this diversification became increasingly popular also with insurance companies. Nevertheless evidence from the literature suggests that bancassurance was -certainly in the beginning- more successful than assurfinance. Insurance companies played a less important role as agents for banking products or were confronted with more threats and drawbacks, not in the least because of the hostile reactions of their intermediaries.

This trend was altered to some extent by the creation of full-fledged financial conglomerates and the more general acceptance of these diversification and multi-channel strategies. Nevertheless a number of differences has to be taken care of, such as the considerable differences on the marketing level and the remuneration structure. Distinction has also to be made between the types of client segments involved. As our market research figured out, this segmentation is to be approached differently from the perspective of banks, insurance companies and insurance intermediaries, diversifying into each other's territorium, hereby taking into account the many divergent views of consumers upon these diversification strategies and marketing approaches.

The all finance trend and its distinction from the cross-selling focus of bancassurance and assurfinance were analysed within the more general perspective of *unbundling and rebundling of services.* This analysis clearly pointed out that the value added of integrated services is potentially much higher than the value created through packaging of services and is certainly superior to mere cross-selling.

The use of the evaluation method developed by Heskett e.a. (1997) helped us in shading the potential advantages and disadvantage of each of these marketing approaches. The results of our market research showed, however, that this evaluation must be differentiated according to the type of client and the complexity of the products involved.

The driving forces behind the all finance trend are to be found in the market environment (competition, deregulation, privatisation of the social security system, etc.). Factors enabling this all finance trend are foremost lying in the IT environment. Also increased use of networking, alliances and the focus on relationship marketing instead of transactional marketing, are valid enablers.

3. Looking for the *ratio behind these diversification strategies* first leads to an in-depth analysis of the diversification literature and its applicability to the all finance movement.

Although diversification can be labelled as one form of a growth strategy, the analysis showed quite diverging rationales for diversification. The most researched diversification motives certainly are synergies and economics of scope. Another stream of literature focuses on the comparison between transaction versus bureaucratic costs in order to justify the make-or-buy, c.q. the diversification, decision. The agency theory focuses on managerial motives like empire building. Other motives include financial synergies and market power. The more studies appeared on the empirical evidence of the relation between diversification and firm performance, the more inconclusive the results became.

Many diversification strategies resulted in bad performance. It seems that managers encounter many difficulties in the process of realising the diversification advantages. The poor track record is also explained by overestimation of the transferability of assets and capabilities and of the competitive power while underestimating the impact of a specific market environment. This led to a revival of the *back to the core business* and *back to basics* strategies, and consequently to de-mergers and refocusing.

In a second step the diversification strategy of financial firms was confronted with the larger business world. At first glance there is an interesting paradox: while financial firms are forming conglomerates, many large firms are refocusing on their core activities, (industrial) organisations are down scaling, refocusing to improve flexibility and are even breaking up (by MBO or otherwise). Our hypothesis is that financial conglomerates are not a *conglomerate* diversification as defined in the (industrial) management literature but mostly a related diversification. As a consequence it is not necessarily a contra-indication of the general business trend. In fact, it seems that the financial sector is moving towards the same types of strategy, but that it is coming from a different background.

4. In order to be able to gain a better insight into the financial conglomerate trend, to detect the differences and similarities, to evaluate the performance and to locate and manage the risks involved, a separate model was developed. This *financial conglomerate control board (F.C.C.B.)* combines the concepts of the value chain and the services production or servuction system. Moreover this model makes it possible to test for the different alternative models developed to structure the financial conglomerates (such as the typologies based on the *corporate structure*, the relative weight of the sectors involved and the method of entry).

   The F.C.C.B. englobes three relevant groups of analysis: the structural factors, the activities -divided into support activities and primary activities- and the environmental factors.

   The level of integration is used as a central reference point in order to classify the financial conglomerates. It's our firm belief that synergies and the risk level are determined by the level of integration and co-operation between the different parts of the group. It is therefore necessary to look more carefully at the concept of integration and related aspects such as organisational autonomy, (de)centralisation and the need for strategic interdependence. It seems that exploiting the synergies is an exercise which requires an optimal balancing between achieving interdependencies between the compounding parts and avoiding the problems of co-ordination and control loss.

5. The F.C.C.B. was used to develop a *comparative analysis of the diversification strategies of 25 Belgian and Dutch financial conglomerates* (which revealed considerable differences between those two countries).

   A. The *structural elements* of each financial conglomerate were analysed at three levels: namely at the legal, the capital and the organisational level. The analysis of the *legal structure* and the organisational structure mainly focuses at the method of entry and shows that there are not only differences between the conglomerates but that even within one conglomerate different methods of entry can be combined. Nevertheless this approach was complemented with a more dynamic analysis: especially alliances between banks and insurance companies often evolved in a sale of the insurer to the bank. The Belgian players seem to opt for a more moderate and less integrated approach towards financial conglomerates compared to their Dutch peers.

The *capital structure*, which highlights the degree of (inter)relations between the partners from a capital perspective forming a financial conglomerate, again showed clear differences between the two countries studied. Belgian conglomerates often opt for less important equity stakes than their Dutch colleagues, hereby accentuating the important differences on the level of governance structures. The reference shareholder system is indeed an important characteristic in Belgium.

The way conglomerates implement their diversification strategies can differ quite substantially. These *organisational structures* have been grouped into 4 broad categories: (1) maintaining the previous structure without fundamental adaptations or links, (2) adding a subsidiary to complement the main (group) company, (3) creating a joint holding company, and (4) creating two joint levels (one at the holding level and one or more joint working companies in between the holding and the different group companies and subsidiaries). This last option, which leads to the most integrated financial conglomerate, is chosen by some of the leaders in this diversification trend (e.g. ING and Fortis).

B. The second part of the F.C.C.B. details the degree of integration of the different activities involved in banking, insurance and other financial services.

For what considers the *support activities* the following conclusions can be drawn.

- The corporate governance and management level englobes all activities in relation to the formation and realisation of the strategy. Large differences exist in practice as to the degree of integration, going from an integrated approach with strategic control over the business units (or subsidiaries) to minimal links and large autonomy to the separate partners.
- The integration of office facilities and other infrastructure seems to be less high on the agenda, even for those groups that integrate on many other levels.
- Most of the integrated groups opt for a centralised investment department or a co-ordination of the ALM while the groups with a less integrated focus have not yet reached or discovered the potential synergies in this respect.

- The information technology still offers enormous potential for further integration (production process, product integration, database marketing, payment systems).
- The more integrated groups become the more emphasis there is (or will be) on the tuning of the respective management information systems and this from a strategic as well as from a reporting perspective.

The analysis of the *primary activities* shows the following results.

- Product development is slowly but certainly becoming more and more co-ordinated; the trend towards modular concepts is gaining importance.
- The integration of the market development could only be studied indirectly through the communication strategy. Although the larger financial conglomerates opt for the variety model (different brand names) they also invest quite substantially in a corporate image. The less integrated the group the lower the level of tuning in communication terms.
- The integration on the level of operations management is especially visible in selling processes of insurance products by banks.
- Most financial conglomerates primarily aim at distribution synergies, although exploiting these synergies has much to do with balancing on a tightrope. The trend points clear to the direction of a more general acceptance of the multi-channel approach.

These findings, although far more detailed and more recent are in line with the results of the study made by Coopers & Lybrand (1993). The main difference lies in the fact that we concluded that it is not sufficient that integration follows the value chain backwards: as long as there is no integration at the top as well, the opportunities and benefits remain restricted. There are just too many difficulties and problems in adequately implementing an all finance strategy. These problems require some form of central co-ordination and management.

6. Although the trend towards financial conglomeration is rather young and the statistics available are not easily available, we tried to *quantify the performance effects of these diversification strategies* in chapter 6. The literature on the performance effects of diversification is abundant but undecisive. We tried to eliminate a number of disturbing factors by focusing only on the financial sector within a limited geographic scope. Furthermore we also integrated strategy implementation with the help of the F.C.C.B. We developed a measure of the degree of integration and tested if

such an integration is a necessary condition for successful diversification.

In order to be able to analyse the diversification effect other relevant performance effects had to be distinguished such as differences in size (economics of scale), geographical diversification and corporate governance structures.

The hypotheses tested were the following:

- financial conglomerates outperform *pure* banks and *pure* insurance companies;
- financial conglomerates with relatively equal insurance and banking parts profit more from the benefits of synergies than asymmetric formulas;
- different types of synergies have different effects on performance and the mode of entry is an important element; organisational fit is as important as strategic fit;
- firm performance is (also) positively related to following control variables: industry profitability, size, internationalisation, stockholders' influence and the risks involved.

Thanks to the support of the Levob Foundation, we compiled an own database on the main European financial institutions, from which we extracted more detailed statistics for a sample of 21 financial conglomerates, 15 specialist banks and 13 specialist insurance companies over the period 1991-95. We used different measures to approach diversification: degree of diversification, mode of diversification and process aspects (based on the F.C.C.B.). We distinguished between financial management, human resources, production and marketing synergies and tested for the degree of integration on each of these levels. Performance measures used included ROA and ROE (before tax) as well as compound sales growth and an efficiency measure.

The results of this research are not very convincing (yet). The theoretical advantages ascribed to diversified financial groups are not directly reflected in their *overall* performance. The group of non-diversifying insurance companies seems to be -for the time being- the most profitable. Caution is however necessary with the interpretation of these results.

- The time frame is too short to make definite conclusions on the real effects of diversification. It takes time to integrate and align strategies; the costs occur much earlier than the benefits.
- The success stories of financial conglomerates on the micro-level are in contrast with the global findings in this research. One explanation for the differences between successful diversification at the micro level and negative performance effects on the sub-industry level could well be that diversification is a difficult strategy that leads to more losers

than winners. This hypothesis is supported by the following results: the performance of highly integrated financial conglomerates is higher than when the banking and insurance subsidiaries strive for autonomy and no co-operation.

This research also revealed some interesting results in relation to other performance drivers.

- The industry in which a firm operates has a major influence on a firm's performance: insurance companies still benefit from a better industry performance. However, given that a real competitive market will lead to a more equal level playing field between banks and insurance companies, it is possible that banks loose their relative disadvantage (from a profitability point of view). If this will be the case, the differential disadvantage of financial conglomerates over specialised insurance companies will probably decrease or disappear.
- The governance structure plays an important role, but contrary to mainstream beliefs the financial sector is better off with managers who take the lead than with strong stockholder influence.
- Banks do better by setting up their own subsidiary, rather than engaging in looser forms of co-operation.
- There are advantages for insurance companies to integrate the investment activities with those of their banking partners.
- Banks realise positive performance effects with marketing integration.

7. In order to *"create the future"* banks, insurance companies and other financial firms will have to build upon the mainstream of *new strategic thinking* but success will depend to a large extent on the *appropriate translation* of these general recipes to the (very) specific environment and characteristics of the financial sector.

   In chapter 7 and 8 attempts were made to develop some specific suggestions and to document on some interesting case studies. Concerning the translation of the strategic recipes the following can be concluded.

   - Although it is clear that *attack is the best of all defences*, following this recipe will have drastic effects in the financial sector which used to opt for a rather defensive strategic course. The days of stable market environments, protected by comfortable and secure (self) regulations are gone. So the market players had to shift gear from a (very) defensive strategic approach to what seems often a very aggressive attack. This is true in the banking as well as in the insurance sector and financial conglomerates are often taking the lead in this respect.

- The shift of the attention from the actual competition and market share struggle towards competing for the future is extremely important for banking and insurance companies. It is therefore of the utmost importance to concentrate upon real improvements for the (potential) clients. Here the route of all finance looks promising for quite an important segment of the market.
- The *role of the client in shaping the future* of the financial sector must not be underestimated. The insurance sector is probably facing the most important challenges in this respect shifting from a technical product orientation towards an integrated client-oriented solution.
- The concept of *stretch* offers interesting perspectives for the financial sector: efficiency can be improved with more long-term success when oriented towards expanding market and product development than when only focusing on BPR. Here again the all finance route offers many opportunities for shifting the vicious circle of price-cutting into a virtuous circle of relationship marketing.
- *Core competencies* are often defined in the insurance and banking practice in relation to technological leadership and customer competencies. The route of networking and alliances in the direction of financial conglomerates are important in this respect but the trend towards financial conglomerates places more emphasis on integration so that alliances and networking are not the main method of diversification but rather a temporary or a complementary route.
- Especially insurance firms with a traditional distribution channel have much to gain from the concept of *leverage* in order to create a learning relationship to improve the efficiency and effectiveness of the service chain. Within financial conglomerates the challenge is to exceed the business unit approach based on technical product characteristics and to integrate (on different levels of the F.C.C.B.) in order to capitalise on the many potential synergies.

The conclusion therefore was that financial conglomerates can create their own future by redefining their core business not in technical terms but from a functional and customer-oriented perspective. Although quantitative research could not reveal significant positive results for every partner involved, the concepts of all finance and financial conglomerates have been used in some cases with spectacular results.

8. We therefore analysed some of these *success stories* into greater detail.

The *ING case* is an example of an integrated financial conglomerate that builds upon all finance or full service concepts. This case shows that far more synergies can be exploited than only cross-selling of each other's products. The F.C.C.B. analysis indeed showed the most diversified use of synergies at all possible levels and this in the home market as well as abroad. Nevertheless the case analysis also made clear that it is a difficult exercise to reach an optimal equilibrium between the different marketing approaches and diverging cultures involved.

The *Fortis case* is another example of a successful creation of *ones own future* . The Fortis case is quite different from the ING case especially for what concerns the legal structure and the integration of the whole group. Nevertheless in its two home countries (linked to its bi-national character) a fully-fledged all finance approach is chosen, while joint ventures are developed abroad. One of their strengths in this diversification process is that they focus on the management of differences and by so doing optimize on these differences (e.g. their bi-national character). Fortis opted primarily for a geographical structure whereas the ING structure is more functionally oriented (closer to the different market segments and functional specialities).

As with most firms, one of the big challenges for the Fortis group is to find an optimal equilibrium between the synergies created by centralisation and the flexibility of decentralisation. Their share price evolution proves that they have been (very) successful (certainly for what concerns the Fortis AG Share).

The *BCP case*, which is another success story, shows fundamental differences compared to the two previous cases. Started as a one-stop banking operation in 1986 in Portugal, it grew into an important (local) financial conglomerate by internal growth as well as by turning minority joint-ventures into integrated group companies. Their three key success factors were: (1) a modern market segmentation, (2) building human resources and (3) forming alliances. All this was achieved with the help of the most modern IT. Their segmentation approach is applied uniformly throughout all of the groups' companies. An important alliance is the one with the Eureko Group, which helps them in realising their internationalisation strategy in the insurance industry.

Besides these three case studies, the development of *insurance derivatives and securitisation* is analysed as another example of pro-active stretch and leverage strategies with high potential for creating ones own future. The all finance movement was primarily limited to the retail market and the collaboration between banks and (direct) insurance companies. These new risk management and financing techniques studied here clearly open up a whole new domain for the exploitation of synergies between primarily investment bankers and (re)insurance companies.

9. The increasing interface between banks, insurance companies and investment institutions not only has high potential for synergies but at the same time alarms the *supervisory authorities* to investigate whether this trend is not also increasing the risk profile of the

companies involved. The *potential risks involved* can be categorised as follows:

- risk of contagion;
- risk of double gearing;
- creation of opaque structures;
- external conflicts of interest;
- internal conflicts of interest and sub-optimality;
- quality downgrading;
- abuse of power.

The *European Commission* wishes to interfere as little as is reasonably possible with this market-driven evolution towards financial conglomerates. They, however, want to maintain the rigorous separation between the credit, investment and insurance activities. This specialisation was also found at the level of the supervisory authorities. Given specific problems that arise in group configurations this solo-approach is gradually expanded with solo-plus supervision and co-ordination between supervisory authorities. Moreover the consolidated supervision of banking groups is giving way to a more group-oriented supervision of insurance companies (chapter 11).

The EC is most concerned with the following:

- the prevention of double gearing;
- excessive risk concentration;
- transparency of group structures;
- fitness and properness of shareholders;
- responsibility of the essential auditors;
- exchange of information between supervisory authorities;
- monitoring of intra-group transactions.

Besides the EC also other supranational organisations like the *OECD*, *IOSCO* and the Basle Committee are working on prudential supervision of financial conglomerates.

In contrast to the European scenery, the *U.S. environment* is much more restrictive towards the diversification of banks and insurance companies. Significant pressure is however exercised especially from the banking sector against these restrictions. Since the passage of the Bank Holding Company Act and the Glass-Steagall Act, baring banks from engaging in securities and insurance operations, the lobby of insurance agents has proven to be very strong for defending the strict specialisation of these business sectors. However signals become visible that a number of leading insurers have begun pulling back from their historical support of these restrictions while the defenders of a more open competition are growing in number and importance.

Moreover large insurance brokers, commercial and reinsurance companies as well as investment banks arrive more and more at a convergence in the U.S. market place. This is partly explained by the growing interface between financial techniques, insurance capacity and risk management (see point 8.4). Growing convergence can also be observed between different types of financial institutions to set up joint ventures to offer specific lines of insurance (catastrophic P & L insurance, credit insurance) to provide investment advice or commercial credit.

10. The *proposal for a European directive on insurance groups* is a necessary step in the direction of a regulation on financial conglomerates. It consists of a number of issues that are also relevant for the handling of all potential risks involved (see chapter 10). We analysed some of them into more detail.

- The analysis of the EC definition of an "insurance group" led to a critique on the strict application of general definition concepts to insurance companies: since insurers are by definition institutional investors, the definitions of *group, parent/subsidiary undertaking and participation* certainly require special attention. There are even considerable differences between the banking and insurance sectors, particularly for what concerns the money-creating potential and the different categories of security capital. The definition criteria were tested in order to have some insight into the possible implication of those different approaches. Especially the threshold to distinguish whether a company forms part of a group is crucial in this respect. The lower the threshold the more companies belong to a group structure but also the more separate groups can be observed (members of a group can act as a parent company themselves, thereby eventually shifting from one sector (e.g. insurance) to another (e.g. banking). Our database showed that there are more group relations within the insurance sector than in the banking industry, which can (partly) be ascribed to the specialisation requirements within the insurance sector.
- The additional regulation on insurance groups (the *plus* in the solo-plus approach) is threefold:
  - improvement of the exchange of information;
  - rules for intra-group transactions;
  - rules to prevent double gearing.

  This last question is studied into greater detail because the definition used by the EC is certainly to be questioned from different points of view. We also tried to test whether the European insurance industry is really using *double gearing* as a capitalisation technique and found that double gearing sensu stricto (cross-shareholdings) does exist especially in countries like France and Belgium. These countries are known for their reference shareholding system, that often uses these cross-shareholding structures. Whereas the banks opted for a consolidated approach in order to avoid double gearing in calculating the solvency capital, the insurance directive proposed two other methods besides this consolidation technique. Although they proclaim that these three methods are equivalent, important divergences certainly exist. The degree in which surpluses within one group company can be used to cope with the insufficient insolvency capital in other group members shows important differences between these three possible methods.

- Finally the question of a solo-plus versus a consolidated supervisory approach has been raised. Due to the growing interface between banks and insurance companies on the one hand and between financial and insurance products on the other hand, market reality seems to be drifting further and further away from the traditional legal division into separate financial sectors and their respective supervisory authorities. It is therefore necessary to reconsider these historical border lines.

REFERENCES

Ackoff, R.L. (1994) *The democratic corporation: a radical prescription for recreating corporate America and rediscovering success*, Oxford University Press, New York.

A.D. Little (1985) Internal documents.

Alexander, M., Campbell, A. and Goold, M. (1995) 'Corporate strategy: the quest for parenting advantage', *Harvard Business Review*, March-April, 120-132.

Amihud, Y. and Lev, B. (1981) 'Risk reduction as a managerial motive for conglomerate mergers', *Bell Journal of Economics*, 12, Autumn, 605-617.

Amit, R. and Livnat, J. (1988) 'Diversification strategies, business cycles and economic performance', *Strategic Management Journal*, 9, 99-110.

Ansoff, H.I. (1965) *Corporate strategy: an analytical approach to business policy for growth and expansion,* McGraw-Hill Book Company, New York.

Argenta *Annual report 1995*, Antwerpen.

Banco Commercial Português *Annual report 1995*, Porto.

Barney, J.B. (1986) 'Strategic factor markets', *Management Science*, 42, 1231-1241.

Barney, J.B. (1991) 'Firm resources and sustained competitive advantage', *Journal of Management,* 17 (1), 99-120.

Bartelds, H. (1996) 'Managing the differences: towards a new generation of financial service providers', *paper presented at the Congress of the International Insurance Society Inc.*, Amsterdam, July.

Berger, P.G. and Ofek, E. (1995) 'Diversification's effect on firm value', *Journal of Financial Economics*, 37, 39-65.

Besanko, D., Dranove, D. and Shanley, M. (1996) *The economics of strategy*, John Wiley & Sons, New York.

Bettis, R.A. and Prahalad, C.K. (1986) 'The Dominant Logic: a new linkage between diversity and performance', *Strategic Management Journal,* 7, 485-501.

Bhagat, S., Shleifer, A. and Vishny, R.W. (1990) 'Hostile takeovers in the 1980s: the return to corporate specialization', *Brookings Papers: Microeconomics*, 1-84.

Blankenhorn, D. (1996) AT&T admits it's Fubar, *Internet WWW, CMP Publications Inc.*

Bleeke, J. and Ernst, D. (1995) 'Is your strategic alliance really a sale?', *Harvard Business Review*, January-February, 97-105.

Bradley, M., Desai, A. and Kim, E.H. (1988) 'Synergistic gains from corporate acquisitions and their division between the stockholders of target and acquiring firms', *Journal of Financial Economics*, 21, 3-40.

Brost, M.A. and Kleiner, B.H. (1995) 'New developments in corporate diversification strategies', *Management Research News*, 18 (3/4/5), 24-33.

Brush, T.H. (1996) 'Predicted change in operational synergy and post-acquisition performance of acquired business', *Strategic Management Journal*, 17, 1-24.

Busija, E.C., O'Neill, H.M. and Zeithaml C.P. (1997) 'Diversification strategy, entry mode, and performance: evidence of choice and constraints', *Strategic Management Journal*, 18 (4), 321-327.

Celarier, M. (1994) 'ING's scrappy emerging markets opportunism', *Global Finance*, December, 78.

Chandler, A.D. Jr. (1962) *Strategy and structure: chapters in the history of the industrial enterprise*, MIT Press, Cambridge; Massachusetts.

Chang, Y. and Thomas H. (1989) 'The impact of diversification strategy on risk-return performance', *Strategic Management Journal*, 10, 271-284.

Chatterjee, S. (1986) 'Types of synergy and economic value: the impact of acquisitions on merging and rival firms', *Strategic Management Journal*, 7 (2), 119-139.

Chatterjee, S. and Wernerfelt B. (1991) The link between resources and type of diver-sification: theory and evidence, *Strategic Management Journal*, 12 (1), 33-48.

Christensen, H.K. and Montgomery, C.A. (1981) 'Corporate economic performance: diversification strategy versus market structure', *Strategic Management Journal*, 2, 327-343.

Collis, D.J. and Montgomery, C.A. (1995) 'Competing on resources: strategy in the 1990', *Harvard Business Review*, July-August, 118-128.

Conner, K.R. (1991) 'A historical comparison of resource-based theory and five schools of thought within industrial organization economics: do we have a new theory of the firm?', *Journal of Management*, 17 (1), 121-154.

Coopers and Lybrand (1993) *Making bancassurance work: survey of European financial institution's policies and practice*, Coopers & Lybrand, European Financial Services Group, Amsterdam.

Coopers and Lybrand (1995) *Ondernemingsanalyses 94/95 - trends bij verzekeringsmaatschappijen*, Delwel Uitgeverij, 's-Gravenhage.

Coopers and Lybrand (1997) *Ondernemingsanalyses 1997 - trends bij verzekeringsmaatschappijen*, Delwel Uitgeverij, 's-Gravenhage.

Coopers and Lybrand (1997) *Ondernemingsanalyses 96/97 - trends bij banken*, Delwel Uitgeverij, 's-Gravenhage.

Courtnage, S. (1994) 'On a roll?', *Money Management*, December, 58-61.

Datta, D.K. and Grant, J. (1990) 'Relationship between type of acquisition, the autonomy given to the acquired firm, and acquisition success: an empirical analysis', *Journal of Management*, 16 (1), 29-44.

Datta, D.K., Rajagopalan, N. and Rasheed, A.M.A. (1991) 'Diversification and performance: critical review and future directions', *Journal of Management Studies*, 28 (5), September, 529-558.

Datta, D.K. (1991) 'Organizational fit and acquisition performance: effects of post-merger integration', *Strategic Management Journal*, 12, 281-297.

Davis, P.S., Robinson, R.B. Jr., Pearce II, J.A. and Park, S.H. (1992) 'Business unit relatedness and performance: a look at the pulp and paper industry', *Strategic Management Journal*, 13, 349-361.

Davis, R. and Thomas, L.G. (1993) 'Direct estimation of synergy: a new approach to the diversity-performance debate', *Management Science*, 39 (11), November, 1334-1346.

De Financieel-Economische Tijd (1995) 'Wall Street juicht opsplitsing van AT&T toe', 21 September.

De Financieel Economische Tijd (1995) 'Opsplitsing AT&T kost 20000 jobs', 28 September, 28 September.

De Financieel-Economische Tijd (1995) 'Brederode bewijst het bestaansrecht van Belgische holdings', 23 September.

De Financieel-Economische Tijd (1997) 'Makelaar wapent zich tegen multidistributie', 22 October.

De Financieel-Economische Tijd (1997) 'Grootste vooruitgang voor Gemeentekrediet en BBL', 27 December.

Dess, G.G. and Origer, N.K. (1987) 'Environment, structure, and consensus in strategy formulation: a conceptual integration', *Academy of Management Review*, 12 (2), 313-330.

Dess, G.G., Gupta, A., Hennart, J.-F. and Hill, C.W.L. (1995) 'Conducting and integrating strategy research at the international, corporate, and business level: issues and directions', *Journal of Management*, 21 (3), 357-393.

Dickinson, G.M. and Dinenis, E. (1994) 'The Changing interface between insurance companies and banks', in *Papers of the 20th Seminar of the European Group of Risk and Insurance Economists (Conference of the Geneva Association)*, Rotterdam, September.

Dubofsky, P. and Varadarajan, P. (1987) 'Diversification and measures of performance: additional empirical evidence', *Academy of Management Journal*, (3), September, 597-608.

Elewaut, G. (1996) 'Het annus horibilis van Suez', in *Dit was 1995*, 67.

Ennew, C., Wong, P. and Wright, M. (1992) 'Organisational structures and the boundaries of the firm: acquisition and divestment in financial services', *The Services Industries Journal*, 12 (4), 478-496.

Fairlamb, D. (1995) 'In search of synergy', *Institutional Investor*, June, 56-60.

Feiner, L. (1995) 'Banque-assurance: mariages forcés', *Trends Tendances*, 19 October, 22-28.

Forbes (1995) 'Big lessons', 16 October.

Forestieri, G. (1993) 'Economies of scale and scope in the financial services industry: A review of recent literature', *Financial Market Trends*, 4, 63-123.

Fortis *Annual report 1995*, Utrecht-Brussel.

Fortune (1995) 'AT&T has the plan', 16 October.

Geringer, J.M., Beamish, P.W. and daCosta, R.C. (1989) 'Diversification strategy and internationalization: implications for MNE performance', *Strategic Management Journal*, 10 (2), 109-119.

Grant, R.M. (1995) *Contemporary strategy analysis: concepts, techniques, applications,* Blackwell Publishers Inc., Oxford.

Gupta, A.K. and Govindarajan, V. (1986) 'Resource sharing among SBUs: strategic antecedents and administrative implications', *Academy of Management Journal*, 29 (4), 695-714.

Hall, E.H., Jr. and St. John, C.H. (1994) 'A methodological note on diversity measurement', *Strategic Management Journal*, 15, 153-168.

Hamel, G. and Prahalad, C.K. (1994) *Competing for the future: breakthrough strategies for seizing control of your industry and creating the markets of tomorrow*, Harvard Business School Press, Boston, Massachusetts.

Hart, C.W.L. (1995) 'Mass customization: basisbegrippen, mogelijkheden en beperkingen', *M & O Quarterly*, 3, 6-21.

Haspeslagh, P.C. and Jemison, D.B. (1991) *Managing acquisitions: creating value through corporate renewal*, The Free Press, New York.

Heffernan, S. (1995) Big Bangs and expansion activity in the financial sector, *Banking Word,* March, 20-22.

Herring, R.J., and Santomero, A.M. (1990) 'The corporate structure of financial conglomerates', *The Journal of Financial Services Research*, 471-497.

Hesberg, D. and Karten, W. (1994) 'Supervision of financial conglomerates: some theoretical remarks', *The Geneva Papers on Risk and Insurance*, 70, 1-21.

Heskett , J.L., Sasser, W.E. and Schlessinger, L.A. (1997) *The service profit chain: how leading companies link profit and growth to loyalty, satisfaction and value*, The Free Press, New York.

Het Verzekeringsblad (1997) 'De pakketten praktisch marktdekkend op een rij', 24 April.

Hill, C.W.L. (1994) 'Diversification and economic performance: bringing structure and corporate management back into the picture', in *Fundamental issues in strategy: a research agenda* (Rumult, R.P., Schendel, D.E. and Teece, D.J. (Eds.)), Harvard Business School Press, Boston, Massachusetts, 247-269.

Hill, C.W.L. and Hoskisson, R.E. (1987) 'Strategy and structure in the multiproduct firm', *Academy of Management Review*, 12 (2), 331-341.

Hill, C.W.L. and Snell, S.A. (1988) 'External control, corporate strategy, and firm performance in research-intensive industries', *Strategic Management Journal*, 9, 577-590.

Hill, C.W.L. and Hansen, G.S. (1991) 'A longitudinal study of the cause and consequences of changes in diversification in the U.S. pharmaceutical industry 1977-86', *Strategic Management Journal*, 12, 187-199.

Hill, C.W.L., Hitt, M.A. and Hoskisson, R.E. (1992) 'Cooperative versus competitive structures in related and unrelated diversified firms', *Organization Science*, 3 (4), November, 501-521.

Hinterhuber, H.H. and Levin, B.M. (1994) 'Strategic networks: the organization of the future', *Long Range Planning*, 27 (3), 43-53.

Holsboer, J.H. (1993) *Specialization and diversification in financial services: some recent, practical experiences in the Netherlands*, Geneva Lecture, Rotterdam, 23 September.

Holsboer, J.H. (1995) 'Only the tip of the iceberg has been discovered', *ING News Journal*, 6, 3.

Hoschka, T.C. (1994) *Bancassurance in Europe*, The MacMillan Press Ltd., Houndmills.

Hoskisson, R.E. and Hitt, M.A. (1990) 'Antecedents and performance outcomes of diversification: a review and critique of theoretical perspectives', *Journal of Management*, 16 (2), 461-509.

Hoskisson, R.E., Hitt, M.A., Johnson, R.A. and Moesel, D.D. (1993) 'Construct validity of an objective (entropy) categorial measure of diversification strategy', *Strategic Management Journal*, 14, 215-235.

Hunt, J.W. (1990) 'Changing pattern of acquisition behaviour in takeover and the consequences for acquisition processes', *Strategic Management Journal*, 11, 69-77.

ING *Annual report 1996*, Amsterdam.

Jacobs, A.G. (1993) 'Evaluating insurance company performance: staying competitive and solvent - the management perspective', in *1993 seminar and proceedings manual - International Insurance Society, Inc.* (International Insurance Society Inc., ed.), International Insurance Society, Inc., 175 -191.

Jaumain, C. (1996) *Financiële gids van de verzekeringen 96*, Kluwer Editorial, Wolters Kluwer België, Diegem.

Jemison, D.B. and Sitkin, S.B. (1986) 'Corporate acquisitions: a process perspective', *Academy of Management Review*, 11 (1), 145-163.

Jensen, M.C. (1986) 'Agency costs of the free cash flow, corporate finance and takeovers', *American Economic Review*, 76, 323-329.

Jensen, M.C. and Ruback, R.S. (1983) 'The market for corporate control: the scientific evidence', *Journal of Financial Economics*, 11, 5-50.

Johnson, G. and Thomas, H. (1987) 'The industry context of strategy, structure and performance', *Strategic Management Journal*, 8 (4), 343-361.

Jones, G.R. and Hill, C.W.L. (1988) 'Transaction cost analysis of strategy-structure choice', *Strategic Management Journal*, 9, 159-172.

Kabir, R., Cantrijn, D. and Jeunink, A. (1997) 'Takeover defenses, ownership structure and stock returns in the Netherlands: an empirical analysis', *Strategic Management Journal*, 18 (2), 97-109.

Kielholz, W. and Durrer A. (1997) 'Insurance derivates and securitization: new hedging perspectives for the US Cat Insurance Market', *The Geneva Papers on Risk and Insurance*, 22 (82), January, 3-16.

Kim, W.C., Hwang, P. and Burgers, W.P. (1989) 'Global diversification strategy and corporate profit performance', *Strategic Management Journal*, 10 (1), 45-57.

Koguchi, K. (1993) 'Financial conglomeration', *Financial Market Trends*, (4), October, 7-62.

Kogut, P. (1988) 'Joint ventures: theoretical and empirical perspectives', *Strategic Management Journal*, 9, 319-332.

Krishnan, H.A., Miller, A. and Judge, W.Q. (1997) 'Diversification and top management team complementarity: is performance improved by merging similar or dissimilar teams?', *Strategic Management Journal*, 18 (5), 361-374.

Kroll, M., Wright, P., Toombs, L. and Leavell, H. (1997) 'Form of control: a critical determinant of acquisition performance and CEO rewards', *Strategic Management Journal*, 18 (2), 85-96.

Kusewitt, J.B., Jr. (1985) 'An exploratory study of strategic acquisition factors relating to performance', *Strategic Management Journal*, 6, 151-169.

Lafferty Business Research (1991) *The allfinanz revolution: winning strategies for the 1990s*, Lafferty Publications Ltd., Dublin.

Lafferty Business Research (1994) *Allfinanz 2000: how retail financial institutions worldwide organise to provide total financial services*, Lafferty Publications Ltd., Dublin.

Lamont, B.T. and Anderson, C.R. (1985) 'Mode of corporate diversification and economic performance', *Academy of Management Journal*, 28 (4), 926-934.

Lloyd, W.P. and Jahera, J.S., Jr. (1994) 'Firm-diversification effects on performance as measured by Tobin's q', *Managerial and Decision Economics*, 15, 259-266.

Lloyd's List (1988) Internal documents, 21 November.

Lubatkin, M. (1987) 'Merger strategies and stockholder value', *Strategic Management Journal*, 8 (1), 39-53.

Mahajan, V. and Wind, Y. (1988) 'Business synergy does not always pay off', *Long Range Planning*, 21 (1), 59-65.
Mahoney, J.T. and Pandian, J.R. (1992) 'The resource-based view within the conversation of strategic management', *Strategic Management Journal*, 13, 363-380.
Maital, S. (1994) *Executive economics: ten essential tools for managers*, The Free Press, New York.
Mariotti, S. and Ricotta, E. (1987) 'Diversification: the European versus the U.S. experience', *Multinational Business*, 1, 23-32.
Markides, C.C. (1995) 'Diversification, restructuring and economic performance', *Strategic Management Journal*, 16, 101-118.
Markides, C.C. and Williamson, P.J. (1994) 'Related diversification, core competences and corporate performance', *Strategic Management Journal*, 15, 149-165.
Mascarenhas, B. (1995) 'International industry evolution patterns', *International Business Review*, 4 (2), 233-246.
Maycock, J. (1986) *Financial conglomerates*, Gower Studies in Finance and Investment, University Press Cambridge.
McDonald, L. (1997) 'Gaining a foothold', *Best's Review P/C,* April, 46-50.
Miller, K.D. and P. Bromiley (1990) 'Strategic risk and corporate performance: an analysis of alternative risk measures', *Academy of Management Journal*, 33 (4), 756-779.
Mitchell, J. (1986) 'Consumer protection' in *Proceedings of the symposium of the European Commission on "Europe & the future of financial services"*, Brussels, 5-7 November, 407-420.
Montgomery, C.A. (1994) 'Corporate diversification', *Journal of Economic Perspectives*, 8 (5), 163-178.
Molyneux, P., Altunbas, Y. and Gardener, E.P.M. (1996) *Efficiency in European banking*, John Wiley & Sons Ltd., Chichester.
Morgan, G. (1994) 'Problems of integration and differentiation in the management of "bancasurance"', *Services Industries Journal*, 14 (2), 153-169.
Morgan, G., Sturdy, A., Daniël, J.P. and Knights, D. (1994) 'Bancassurance in Britain and France: innovating strategies in the financial services', *The Geneva Papers on Risk and Insurance*, 19 (71), 178-195.
Moss Kanter, E. (1994) 'Collaborative advantage: the art of alliances', *Harvard Business Review*, July-August, 96-108.
Mueller, D.C. (1969) 'A theory of conglomerate mergers', *Quarterly Journal of Economics*, 83, November, 643-659.
Nathanson, D.A. and Cassano, J. (1982) 'What happens to profits when a company diversifies?', *Wharton Magazine*, Summer, 19-13.
Nayyar, P.R. (1992) 'On the measurement of corporate diversification strategy', *Strategic Mangement Journal*, 13, 219-235.
Nguyen, T.H., Séror, A. and Devinney, T.M. (1990) 'Diversification strategy and performance in Canadian manufacturing firms', *Strategic Management Journal*, 11, 411-418.

Paez, B.L., Sanders, L. and Williams, J.R. (1988) 'Conglomerates revisited', *Strategic Management Journal,* 9, 403-414.

Pekar, P. (1988) 'A strategic approach to diversification', *Journal of Business Strategy,* 5, 99-104.

Penrose, E.T. (1959) *The theory of the growth of the firm,* Basil Blackwell, Oxford.

Peyrelevade, J. (1993) 'La bancassurance, à la fortune du mot', *The Geneva Papers on Risk and Insurance,* 69, 399-402.

Pine II, B.J., Peppers, D. and Rogers, M. (1995) 'Do you want to keep your customers forever?', *Harvard Business Review,* March-April, 103-114.

Pitts, R.A. (1980) 'Towards a contingency theory of multibusiness organization design', *Academy of Management Review,* 5 (2), 203-210.

Porter, M.E. (1985) *Competitive advantage: creating and sustaining superior performance,* The Free Press, New York.

Porter, M.E. (1980) *Competitive strategy, techniques for analysing industries and competitors,* The Free Press, New York.

Porter, M.E. (1987) 'From a competitive advantage to corporate strategy', *Harvard Business Review,* 63, 43-59.

Priem, R.L. (1990) 'Top management team group factors, consensus, and firm performance', *Strategic Mangement Journal,* 11, 469-478.

Ramanujam, V. and Varadarajan, P. (1989) 'Research on corporate diversification: a synthesis', *Strategic Management Journal,* 10 (6), 523-551.

Riordan, M.H. and Williamson, O.E. (1985) 'Asset specificity and economic organization', *International Journal of Industrial Organization,* 3, 365-378.

Robens, H. (1987) 'Ursachen und Folgen des Wandels der Wachstungstrategie bei den Anbietern Finanzieller Dienstleistungen', *Versicheringswirtshaft,* 2, 106-112.

Robins, J. and Wiersema, M.F. (1995) 'A resource-based approach to the multibusiness firm: emperical analysis of portfolio interrelationships and corporate financial performance', *Strategic Management Journal,* 16, 277-299.

Roquebert, J.A., Phillips, R.L. and Westfall, P.A. (1996) 'Markets vs. management: what drives profitability?', *Strategic Management Journal,* 17, 653-664.

Rowe, W.G. and Wright, P.M. (1997) 'Related and unrelated diversification and their effect on human resource management controls', *Strategic Management Journal,* (4), 329-338.

Rumelt, R.P. (1974) *Strategy, structure and economic performance,* Graduate School of Business, Harvard University, Boston, MA.

Rumelt, R.P. (1982) 'Diversification strategy and profitability', *Strategic Management Journal,* 3, 359-369.

Rumelt, R.P. (1991) 'How much does industry matter?', *Strategic Management Journal,* 12 (3), 167-185.

Sambharya, R.B. (1995) 'The combined effect of international diversification and product diversification strategies on the

performance of US-based multinational corporations', *Management International Review*, 35 (3), 197-218.
Saunders, A. and Walter, I. (1994) *Universal banking in the United States: What could we gain? What could we lose?*, Oxford University Press, New York.
Schwebler, R. (1989) 'Geldvermögensbildung aus der Sicht der Versicherungswirtschaft', *Versicherungswirtschaft*, (16), 1030-1036.
Servaes, H. (1996) 'The value of diversification during the conglomerate merger wave', *Journal of Finance*, 51 (4), September, 1030-1225.
Seth, A. (1990) 'Value creation in acquisitions: a reexamination of performance issues', *Strategic Management Journal*, 11, 99-115.
Shelton, L.M. (1988) 'Strategic business fits and corporate acquisition: empirical evidence', *Strategic Management Journal*, 9, 279-287.
Shimpi, P. (1997) 'The context for trading insurance risks', *The Geneva Papers on Risk and Insurance*, 22 (82), January, 17-25.
Shleifer, A. and Vishny, R.W. (1991) 'Takeovers in the '60s and the '80s: evidence and implications', *Strategic Management Journal*, 12, 51-59.
Shrivastava, P. (1986) 'Postmerger integration', *Journal of Business Strategy*, 7, 65-76.
Simmonds, P.G. (1990) 'The combined diversification breadth and mode dimensions and the performance of large diversified firms', *Strategic Management Journal*, 11, 399-410.
Smith, R.E., Canelo, E.A. and Mi Dio, A.M. (1997) 'Reinventing reinsurance using the capital markets', *The Geneva Papers on Risk and Insurance*, 22 (82), January, 26-37.
Stafford, E.R. (1994) 'Using co-operative strategies to make alliances work', *Long Range Planning*, 27 (3), 64-74.
Swiss Re (1991) 'Economies of scale in the insurance industry', *Sigma, Swiss Reinsurance Company*, (4), 1-20.
Swiss Re (1992) 'Bancassurance: a survey of competition between banking and insurance', *Sigma, Swiss Reinsurance Company*, (2), 1-26.
Swiss Re (1996) 'Insurance derivatives and securitization: new hedging perspectives for the US catastrophe insurance market', *Sigma, Swiss Reinsurance Company*, (5), 1-24.
Teece, D.J. (1980) 'Economies of scope and the scope of the entreprise', *Journal of Economic Behavior and Organization*, 1, 223-247.
The Economist (1990) 'Banking brief: dangerous liaisons', 317 (7677) 20 October, 102-105.
The Financial Times (1996) 'Demerging British Gas', 7 February.
Tripartite Group of Banks (1995) *Securities and Insurance Regulators: The supervision of financial conglomerates*, July.
Vancil, R.F. (1979) *Decentralization: managerial ambiguity by design*, Dow Jones-Irwin, New York.
Van den Berghe, L.A.A. (1989) 'The financial service revolution and the diversification in the financial and insurance world', in

Conference papers on 'Globalisation and Financial Services Revolution: Threat or Opportunity for Banks and Insurance Companies, Brussels, March.

Van den Berghe, L.A.A. and Verweire, K. (1992-1993) *Onderzoek in de consumentenmarkt naar houdingen en gedragingen ten aanzien van risico, sparen, beleggen en verzekeren,* De Vlerick School voor Management.

Van den Berghe, L.A.A. (1994) 'Financiële conglomeraten: strategie op de grens van het (on)haalbare?', *Holland Management Review*, 38, 35-39.

Van den Berghe, L.A.A. (1994) 'Financial conglomerates: opportunity or threat?', in *Risiko, Versicherung, Markt - Festschrift für Walter Karten zur Vollendung des 60. Lebensjahres* (Hesberg, D., Nell, M. and Schott, W., eds.), Verlag Versicherungswirtschaft, Karlsruhe, 445-462.

Van den Berghe, L.A.A. and Baeten, X. (1994-1995) *Survey with regard to the management of an insurance broker's office,* Research Project, De Vlerick School voor Management.

Van den Berghe, L.A.A. (ed.) (1995) *Financial conglomerates: new rules for new players?*, Kluwer Academic Publishers, Dordrecht.

Van den Berghe, L.A.A. and Baeten, X. (1995-1996) *Risico, beleggen, sparen, verzekeren, zorg; Survey in the Dutch Market,* Research Project, Erasmus Universiteit Rotterdam - De Vlerick School voor Management.

Van den Berghe, L.A.A. (1996) 'Der Wettlauf um die Zukunft in der Bank- und Versicherungsbranche: Welche strategischen Konzepte und Erkenntnisse sind nützlich oder sogar notwendig?', in *Festschrift Heinrich Stremitzer*, Wirtschaftsuniversität Wien, 1-13.

Van den Berghe, L.A.A. (ed.) (1996) *From bancassurance and assurfinance towards all finance*, Conference Papers, 19 and 20 March, Brussels.

Vander Vennet, R. (1994) *Determinants of EC bank takeovers and their performance effects: an empirical analysis 1988-1992*, Unpublished doctoral dissertation, University of Ghent.

Vander Vennet, R. (1996) 'The effect of mergers and acquisitions on the efficiency and profitability of EC credit institutions', *Journal of Banking and Finance*, 20, 1531-1558.

van Riel, C.B.M. (1995) 'Corporate communication in financiële conglomeraten', *Holland Management Review,* 42, 60 -68.

van Rüth, V. (1993) *Kooperation und Konzernbildung als Diversifikationsstrategien von Banken in den Versicherungsbereich: Ein Modell zur Vorauswahl, Gestaltung und Bewertung alternativer Diversifikationskonzepte aus Sicht der Kreditwirtschaft*, M&P Verlag für Wissenschaft und Forschung, Stuttgart.

Venkatraman, N. and Ramanujam, V. (1986) 'Measurement of business performance in strategy research: a comparison of approaches', *Academy of Management Review*, 11 (4), 801-814.

Very, P. (1993) 'Success in diversification: building on core competences', *Long Range Planning*, 26 (5), 80-92.

Wernerfelt, B. and Montgomery, C.A. (1988) 'Tobin's q and the importance of focus in firm performance', *American Economic Review*, 78 (1), March, 246-250.

West, C.T., Jr. and Schwenk, C.R. (1996) 'Top management team strategic consensus, demographic homogeneity and firm performance: a report of resounding nonfindings', *Strategic Management Journal*, 17, 571-576.

Williams, K.L. (1986) *Direct marketing of consumer insurance to bank customers,* Executive Entreprise Publications C° Inc., New York.

Williamsion, O.E. (1975) *Markets and hierarchies: analysis and antitrust implications*, The Free Press, New York.

Wrigley, L. (1970) *Divisional autonomy and diversification*, Unpublished doctoral dissertation, Harvard Business School.

INDEX

MIX
Papier aus verantwortungsvollen Quellen
Paper from responsible sources
FSC® C105338

If you have any concerns about our products,
you can contact us on
ProductSafety@springernature.com

In case Publisher is established outside the EU,
the EU authorized representative is:
Springer Nature Customer Service Center GmbH
Europaplatz 3, 69115 Heidelberg, Germany

Printed by Libri Plureos GmbH
in Hamburg, Germany